Trim Carpentry

TAUNTON'S

FOR PROS BY PROS®

BUILDER-TESTED | CODE APPROVED

Trim Carpentry

FROM THE EDITORS OF

Fine Homebuilding

The Taunton Press

The Taunton Press, Inc., 63 South Main Street, PO Box 5506, Newtown, CT 06470-5506
e-mail: tp@taunton.com

Editor: Alex Giannini
Copy editor: Seth Reichgott
Indexer: Jay Kreider
Interior design: Carol Singer
Layout: Rita Sowins/Sowins Design
Cover photographers: (Front cover): John Ross; (Back cover): James Kidd

LIBRARY OF CONGRESS CATALOGING-IN-PUBLICATION DATA

Trim carpentry / from the editors of Fine homebuilding.
p. cm.
Includes index.
ISBN 978-1-60085-502-3
1. Trim carpentry. I. Fine homebuilding.
TH5696.T697 2012
694'.6--dc23
2011050933

PRINTED IN THE UNITED STATES OF AMERICA
10 9 8 7 6 5 4 3 2 1

Homebuilding is inherently dangerous. From accidents with power tools to falls from ladders, scaffolds, and roofs, builders risk serious injury and even death. We try to promote safe work habits throughout this book, but what is safe for one person under certain circumstances may not be safe for you under different circumstances. So don't try anything you learn about here (or elsewhere) unless you're certain that it is safe for you. If something about an operation doesn't feel right, don't do it. Look for another way. Please keep safety foremost in your mind whenever you're working.

ACKNOWLEDGMENTS

Special thanks to the authors, editors, art directors, copy editors, and other staff members of *Fine Homebuilding* who contributed to the development of articles in this book.

Contents

PART FOUR: PANELING

INTRODUCTION

For over 30 years the editors of Fine Homebuilding magazine have been pondering one simple question: What qualities define a "fine" home? Some things are tough to argue against: The importance of the foundation and framing, the long service of durable materials, indoor air quality, and energy-efficiency always make the top of the list. Our readers respond in turn and consistently rank these topics among the most important to their work and their homes. Though it doesn't keep your house standing, make your paint last longer, keep you healthy, lower your monthly utilities bills, or contribute in any significant way to global warming, there is one more topic that the editors here at Fine Homebuilding and our readers, too, find to be a defining element in a high-quality home: Trim.

As builders, we do what we must to get the job done. We may be roofing one week and insulating the next, but we're all trim carpenters at heart. So as soon as the last coat of drywall compound is dry and sanded, we hit the lumberyard, roll out our miter saws, sharpen our pencils, and begin to add the details that define our work. As we case windows and doors, run baseboard and crown, install wainscot and coffered ceilings, the house becomes a home. While it may be fun, installing trim is not easy, mind you. It is skilled work. And work that you or your clients will have to look at for years to come. It's important to get the proportions right, the joints tight, and the transitions as smooth as possible. Here's where this book fits into your tool box.

This collection of trim carpentry projects is written by the country's best trim carpenters and edited by a bunch of trim carpenters at heart. The collective experience found in this book is not only inspirational, it just may help you avoid the rookie mistakes that many of us make on our first projects, or when we attempt to execute a detail for the first time. At *Fine Homebuilding* we've been training trim carpenters for three decades because trim is the most visible mark of a "fine" home. Here are some of our best trim stories from the past few years.

Build well,

Brian Pontolilo, editor, *Fine Homebuilding*

PART 1

Basics

8 Basic Rules to Master Trim Carpentry

BY TUCKER WINDOVER

I'm a busy contractor with a half-dozen trim carpenters working on two or three jobs on any given day. Needless to say, I've had a number of employees on my crew of carpenters over the years. To maintain quality and consistency, I've written up a list of work habits and procedures that I've organized in a three-ring binder. On their first day, every new carpenter receives a copy. It's as much a list of results as a list of techniques. Each point sounds minor by itself, but added together, this list creates a foundation for efficient finish carpentry. Even veteran carpenters can let these simple guidelines slip away from them over time, and that can result in careless work. Finish carpentry is more than just tight miter joints. It's a method of work defined by standards that can be easily replicated.

When the job site becomes a mess, people no longer care.

EXCERPTED FROM A CONTRACTOR'S NEW-EMPLOYEE HANDBOOK, these fundamentals will improve the efficiency and quality of any finish carpenter's work.

1. Keep the site clean

I knew a guy everyone called Yard Sale because he left tools all over the job site. He could never put his hands on the tools he needed. To set up an efficient site, keep tools organized, plan tasks for simple repetition, and lay out job-site materials so that they are easy to access and find.

CORRAL STAIR PARTS AT THE STAIRS. These smaller pieces are easy to misplace, so put everything within easy reach of the stairs. The risers go in one pile, with treads nearby and newel posts next to the handrail fittings.

IT'S BETTER TO HAVE SOME MATERIAL IN EACH ROOM RATHER THAN TO HAVE ONE BIG PILE. Start each project by unpacking the doors and setting them at the openings where they will be installed. This is the time to double-check that the doors swing correctly, that the doors don't swing over a light switch, and that they open cleanly against walls. Stock enough window casing, door trim, and baseboard in each room to complete that room. This way, it's much easier to account for missing material.

ALWAYS USE A NAIL SET. In an average 2,800-sq.-ft. house, there might be 5,000 nails in the trimwork. The painters shouldn't find one unset nail after the finish carpentry is complete.

Don't expect the painter to make your work look good.

2. Be neat about nailing

One good work habit to develop is establishing a pattern to your nailing. On standing and running trim, place nails regularly in pairs every 16 in., or until the material is tight. This keeps the work neat and orderly. Avoid nails inside molding profiles; it's hard for a painter to fill and sand these holes. Use 2-in. nails for baseboard because 2½-in. nails will eventually hit a wire in a 2×4 wall. Use 18-ga. nails for wood-to-wood connections and 15-ga. nails for applying molding over drywall and for setting doors.

AFTER THE CARPENTERS ARE GONE, NO ONE INSPECTS THE HOUSE WITH A LEVEL. Converging lines, however, will stand out to anyone with a good eye. This is especially true where a door is set close to a wall, as in a hallway. If the wall is out of plumb and the door is hung plumb, the casing will show a taper against the wall. However, the aesthetics must be balanced with the door's function, especially if a door is likely to be kept open. Hung out of plumb, a door may swing shut by itself. In that case, the door must be installed so that it operates properly.

Parallel trumps plumb installations, usually.

3. Think and see straight and parallel

Installing most trim is the discipline of connecting two points with a straight line. At this stage in the building process, the work sometimes becomes more a game of appearances than of perfection. The trick is to make bows, bends, and out-of-plumb conditions appear straight and true.

SNAP A LINE AS A GUIDE for the bottom edge of crown molding. If there are waves in the ceiling, at least the bottom edge will be true. Thin gaps at the ceiling can be caulked, and wider gaps can be reduced by flexing the crown to conform to the ceiling.

4. Improve the surfaces before the finish goes on

It may not be noticeable now, but after semigloss paint hits the trim, any sawblade marks, tearout, or imperfections will stand out like a sore thumb. Carry a piece of 150 grit sandpaper in your tool belt. On wood that gets a clear finish, erase or lightly sand out pencil marks. Sand field joints so that they become flush. When you walk away from the work, the stock should be ready for paint or stain.

TIP: A rag dipped in denatured alcohol can erase many pencil marks and doesn't affect the surface of the wood.

6. Know when to make precise cuts

Not every joint shows, so don't waste valuable time on cuts that don't matter. There are plenty of times when a miter or cope will be covered with a successive layer of trim, so only the visible part of the joint needs to be tight. Corners are almost never square, so miters cut at 45 degrees (see the top right photo) will often show a gap. Instead, use a back-beveled miter that creates a tight outside joint; the gap (see the bottom right photo) will be hidden by the band molding. Likewise, shoe molding will cover gaps under a run of baseboard. When installing base or crown at an inside corner, the first piece can be cut a little short because the small gap will be covered by the cope of the next piece.

5. Make miters flush

After door and window casings are painted, mating surfaces that aren't flush will stand out. A miter should be tight and flush, but a thin gap in the miter can be filled with caulk and will disappear. Uneven surfaces will stand out and should be leveled. A shim placed under one side of the miter (see the top photo) can help to line up the joint. Tack it in place, and trim the excess. A little sanding can blend discrepancies between the two sides.

TIP: Use a small combination square to mark a consistent reveal on standing trim.

7. Always use a reveal on standing trim

Avoid a flush joint on layers of standing trim, such as applied window stool or door casings. Typically, any additional layer of trim on a door or window jamb should have a reveal, or it will leave a distracting seam. The size of the reveal depends on the proportions of the trim, but a 3⁄16-in. reveal is a good rule of thumb.

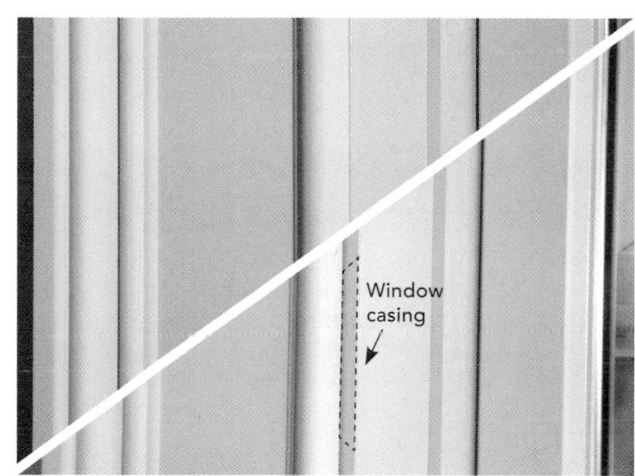

8. You don't have to pull out your tape every time you mark and measure

Most trim measurements can be made faster and more accurately with a more appropriate technique. Choose the most appropriate method for the task at hand.

Pinch sticks

If you can't see the joint, don't waste time on it.

FOR REPEATED MEASUREMENTS OR A SERIES OF MEASUREMENTS taken from the same reference, use a story pole. They are typically made from scrap at hand and have the room to display measurements legibly.

MARK A PIECE OF STOCK IN PLACE. After cutting one end to fit, locate the stock, and use a knife or sharp pencil to scribe the cutline.

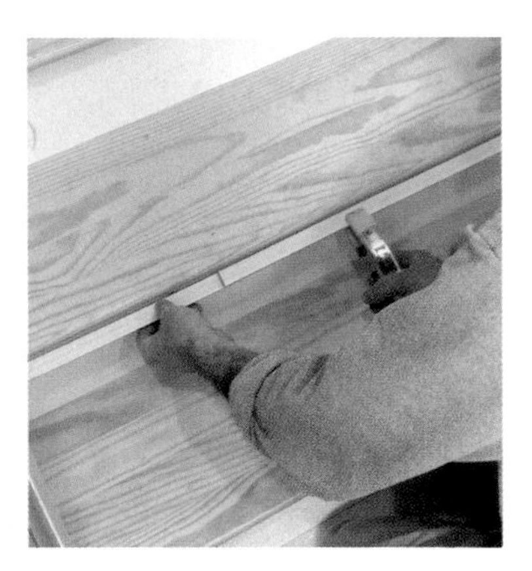

PINCH STICKS LET YOU TAKE AN EXACT MEASUREMENT between surfaces and transfer it to the stock. The simplest version consists of identical rips of ¾-in. stock that are extended and locked in position with a spring clamp. Bevel the outer stick ends for a more accurate read.

To measure any angle, cut a scrap piece of the stock, hold it in place, and mark the outside and inside lines.

Next, place the stock on the opposite side, and mark the lines.

Transfer the crossed marks onto the scrap, and use it to determine the angle of the cut.

Beautiful Trim from Plywood

BY MICHAEL STANDISH

Prices are up, but hardwood-lumber quality is down. If you're disappointed with this state of affairs but still want good-looking trim in your house, you might want to investigate engineered products, such as veneer plywood.

Because of its balanced, cross-ply construction (that is, thin layers of wood with lots of glue to restrain them), veneer plywood is not susceptible to the mutating forces that bedevil ordinary lumber.

Genuinely warped or twisted plywood is difficult to produce unless you leave a sheet of it out in the rain. Swaybacked plywood, resulting from careless storage practices such as deep stacks or inadequate support, is corrected easily by slight hand pressure during assembly. If need be, any unwanted set can be reversed with the process of strategic clamping, or "training" it back to flat.

Finally, veneer-core plywood is stable in another important way. It is far less sensitive than solid wood to fluctuations in humidity, and it expands or contracts only a third as much as Honduras mahogany, a species notorious for staying put.

Plywood also is significantly less expensive. At current local prices, using solid material instead of paint-grade birch or maple plywood means tripling my materials costs, including veneer edging. Hardwood lumber costs me at least twice as much as an equivalent amount of same-species plywood.

But using plywood is not all roses

To be fair, however, at about 75 lb., a ¾-in.-thick, 4×8 plywood sheet is bulky and heavy. Breaking down a full panel with a circular saw or a tablesaw without side and outfeed support can be awkward and slow.

There also are indirect costs. You may have to purchase full sheets, forcing you to store more material than you want. Some suppliers sell partial sheets, but this seeming benefit usually comes with a surcharge, which can be hefty. Also, lengths beyond 8 ft. typically are a special-order item, and any savings gained in materials tends to evaporate.

Another downside is that face veneers are normally no thicker than 1⁄32 in., making veneer-plywood trim too delicate for use in barrooms or fraternity houses. Where trim work will receive merely normal levels of abuse (e.g., outside baseboard corners), veneered material is still not the best choice. For other trim applications such as window and door trim or moldings that are located off the floor, however, this fragility doesn't come into play.

(Continued on p. 17)

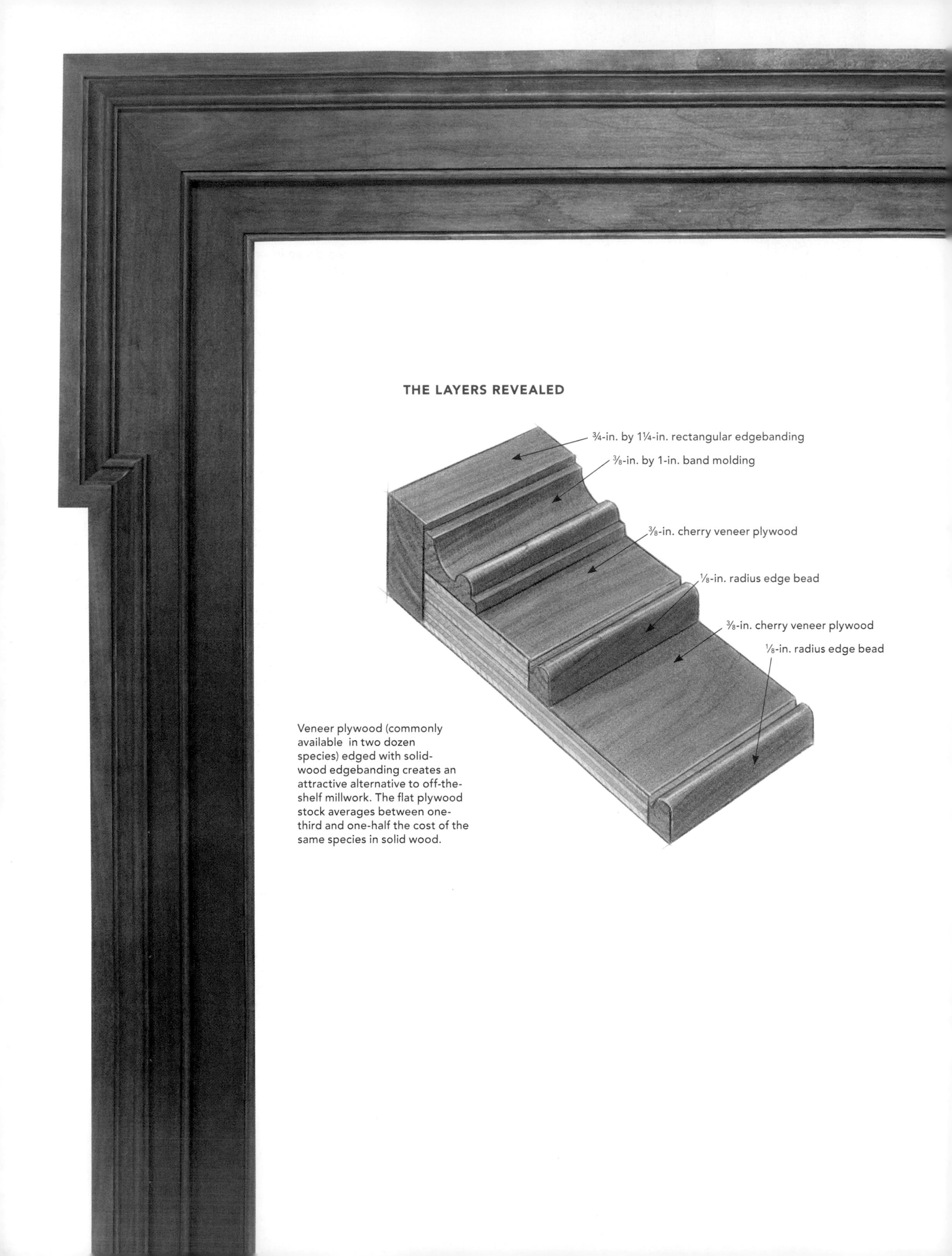

Veneer plywood (commonly available in two dozen species) edged with solid-wood edgebanding creates an attractive alternative to off-the-shelf millwork. The flat plywood stock averages between one-third and one-half the cost of the same species in solid wood.

Solid-wood
molding used as
edgebanding

Square-edged
solid wood used
as edgebanding

Mitered
plywood return

Solid-wood molding
used as edgebanding

THREE WAYS TO HIDE THE EDGES

WHEN USING PLYWOOD AS TRIM, you'll need to cover the edges in most cases. You can do so in one of three ways: solid-wood edgebanding (square-edged stock or molding), a mitered plywood return, or veneer edgebanding (pressure-sensitive or iron-on).

1 Solid-wood edgebanding can be glued on, nailed on, or both to cover plywood's visible layers. A ¼-in.-thick solid-maple nosing was added to the window stool in the photo to protect the vulnerable edge. To protect the plywood's thin veneer, plane the edgebanding down to the tape, then switch to sandpaper.

2 A mitered return cut from the same piece of plywood stock eliminates color-matching problems when staining. Small returns such as this one are often just glued in place.

3 Iron-on veneer edgebanding must contact the plywood edge completely for thorough adhesion.

4 Clean-up is best with a file (10-in. to 12-in. mill bastard) held at about 5 degrees; veneer saws are clumsy, and knives sometimes follow the grain.

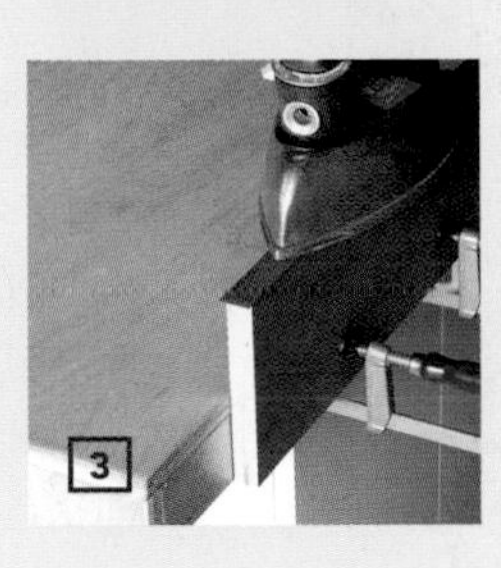

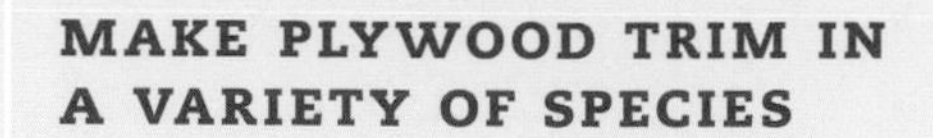

MAKE PLYWOOD TRIM IN A VARIETY OF SPECIES

Because they're readily available, common or exotic species of plywood can be substituted for solid wood. Below are three window-trim examples, from simple to more complex, in pine, mahogany, and oak.

MARKER TIP When only a narrow plywood edge is visible, a brown marker can color the exposed edge and save you the trouble of banding the edge.

PINE

MAHOGANY

OAK

STORE-BOUGHT MOLDINGS CREATE SOLID-WOOD EDGEBANDING
3⁄16-in. by 1¼-in. square-edged solid pine

5⁄8-in. by 1-in. base cap

¾-in. by 4½-in. pine veneer plywood

3⁄8-in. by ¾-in. glass bead

VENEER EDGEBANDING WITH CORNER BLOCK
¾-in. by 4½-in. veneered mahogany casing

1-in. by 5-in. by 5-in. solid mahogany corner block

VENEER EDGEBANDING
Joints are cut square and butted.

The ¾-in. by 4½-in. vertical casing supports a thicker 1-in. by 5-in. top casing, which is made of ¾-in. veneer plywood backed by an additional layer of ¼-in. plywood.

CREATE CUSTOM MOLDINGS WITH A ROUTER

[1] If you can't find the molding you want at your local home center or lumberyard, consider making your own. You can create complex moldings by assembling simple profiles milled with a handheld router or router table. The three router bits below are available from Lee Valley® (www.leevalley.com).

[2] Although designed to make base-cap molding, the bit pictured is similar to the one used to create the band molding for the pine sample on the facing page.

[3] The two bits pictured were used in the more ornate cherry casing on p. 14.

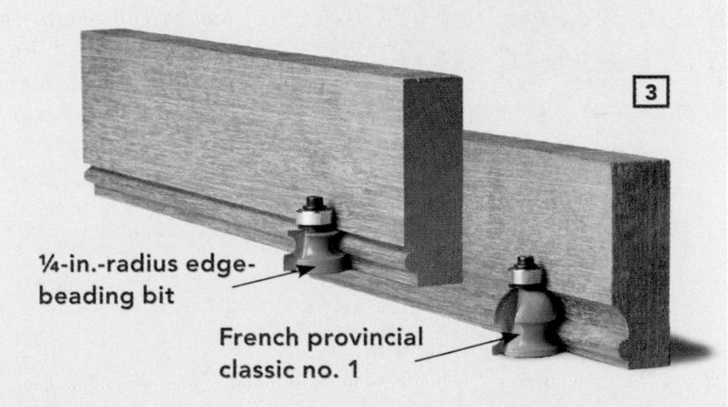

ASSEMBLE THE TRIM FIRST, THEN CUT IT ONCE

Rather than running six individual pieces of trim around a door or window, you can glue, align, and clamp them to create a composite molding, which you then can miter as a single piece. To avoid dangerous shrapnel and to protect the sawblade, be careful not to cut through any nails.

Finally, unless striped edges arc a design element, they must be concealed. It takes time to apply even the simplest banding, such as iron-on or pressure-sensitive veneer tape, but not much more than removing saw marks from lumber.

Wood moldings such as backbanding, either plain or with a profile, are applied to the edges of plywood in much the same manner as solid wood. These moldings, either store bought or made with routers and tablesaws, are nailed and sometimes glued to stable plywood edges.

Paint-Grade Interior Trim

BY CHRIS ERMIDES

If you're a *Fine Homebuilding* reader, you probably love wood. Like it or not, though, when it comes to painted trim, the alternatives to solid wood can make more sense. Finger-jointed, medium-density-fiberboard (MDF), and synthetic trim come with primer coats applied at the factory, which saves you a load of time and trouble. But choosing among these three materials isn't easy.

Finger-jointed trim is knot-free, but it can be pricey. MDF is inexpensive, but it can't get wet. Synthetics come in complex profiles, but they require special adhesives. The look you're after and the specific application (baseboard, casing, crown, or chair rail, for example) will figure in your selection, along with cost, workability, durability, and dimensional stability.

Moldings have a purpose

According to Brent Hull, a historic-moldings expert, interior moldings do two things. First, they define the major shapes and spaces in a room, like doorways, windows, and floor-to-wall and ceiling-to-wall connections.

The second thing moldings do gets at the heart of how to tell good-quality from poor-quality trim. When trim is on a wall, it refines a room architecturally. Profile design, detailing, and scale create the architectural style and tone of the space. But the visual impact is often a subtle, subconscious thing you might not immediately recognize. That visual impact is largely a function of how well-defined the profile shapes are and how sharp the profile's edges are. Hull pushes this point further, adding, "When you look at moldings once they're installed, you should be able to easily read the shapes that make up the profile."

Veteran trim carpenter and *FHB* contributing editor Gary M. Katz agrees: "Crisp, sharp edges create crisp, sharp shadowlines. The relationship between shadow and light is what defines an attractive molding profile, one that can be seen and enjoyed from up close or from a great distance."

Sharp edges should form around transitions between the shapes that make up the profile. This is especially important with crown molding because it's installed overhead and because light doesn't hit it directly. If the profile is muddied with shapes that aren't deeply, crisply cut, it will look like fuzzy lines against the ceiling.

It's important to note, though, that the quality of a molding's profile isn't a function of the material it's made from. You can find sharp or muddy profiles on moldings made of finger-jointed wood, MDF, or synthetics.

Finger-jointed trim is more stable than solid wood

Finger-jointed trim gets its name from the interlocking joint used to connect short boards together end to end. Typically made from various species of pine, finger-jointed trim is free of knots and other defects often associated with solid wood. It's also less likely to cup because finger-jointed trim wider than 6 in. is made by edge-laminating narrower sections together, as shown in the photos on p. 23.

In the past, carpenters complained that finger-jointed trim fell apart at joints or that joints telegraphed through finished surfaces. But according to the manufacturers and expert installers that I talked to, joints fail or telegraph because of poor quality control during the manufacturing process, not because of an inherent weakness in all finger joints.

Kevin Platte, general manager of operations at Windsor Mill in Willits, Calif., said that a strong finger joint depends on two things: tight-fitting fingers and high-quality glue. Although he wouldn't specify the glue Windsor Mill uses, he did say that they, as well as most manufacturers, use a polyvinyl acetate (PVA) glue. For them, using an ASTM-tested glue that creates a joint stronger than the wood is a must. They and others fortify the glue to make it waterproof.

When a finger joint telegraphs, or shows through the finish, it's often because wood from different-age or different-species trees is joined. If adjacent blocks in a finger-jointed blank expand and contract at significantly different rates, joint-telegraphing is likely. Trim made with wood of the same species and with trees of the same age is more uniform and stable.

Sharp knives and slower feed rates yield crisp, consistent profiles

One nice thing about finger-jointed trim is that it can be milled with crisply defined profiles. Like solid-wood molding, finger-jointed molding is shaped in powerful machines equipped with molding heads that hold profiled steel knives.

PROFILE DEFINITION IS THE FIRST SIGN OF QUALITY

WHEN BUYING TRIM, look for clearly defined shapes, deep contours, and sharp edges. Profiles with these characteristics yield correspondingly fine shadowlines, creating an overall impression of refinement and good craftsmanship. Subtle curves with soft edges create fuzzy shadowlines, which make profile shapes hard to read once the trim is installed.

CROWN

Soft edges become even softer once installed. Uninterrupted by furniture or fixtures, crown molding can do more to unify a room than any other molding treatment. That's why sharp edges and deep, well-defined profiles make such a big difference. The subtle transitions between shapes and soft edges found on lower-quality crown contribute little visual impact.

Profiles shown: WindsorONE® Classical Craftsman; generic 3⅝-in. colonial.

CASING

Shapes should be clearly defined. Less-expensive casing lacks the depth necessary to create definite shapes. High-end versions are cut distinctly and are thick enough to stand proud of the baseboard.

Profiles shown: WindsorONE Colonial Revival; generic 3½-in. colonial.

BASE

Thicker stock runs straight. Base moldings milled from thick stock have deeper, more-pronounced details and are rigid enough for straighter runs. Because it bends easily, thin base highlights every subtle wave in the wall.

Profiles shown: WindsorONE Greek Revival; generic 5¼-in. speed base.

Crown

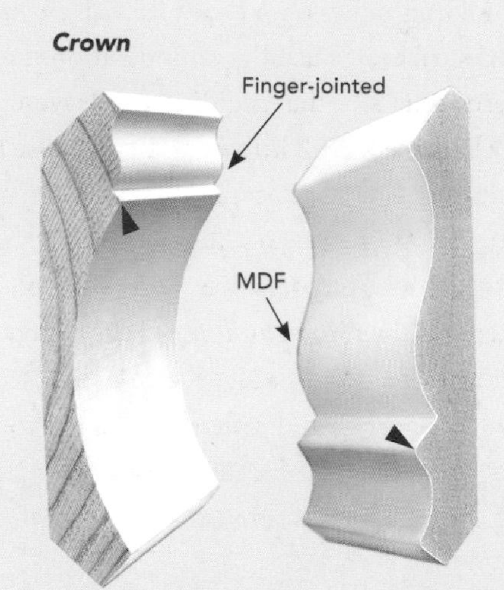

Casing

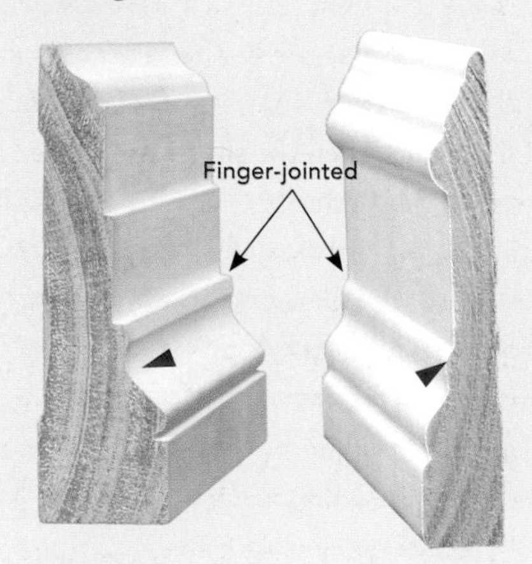

Baseboard

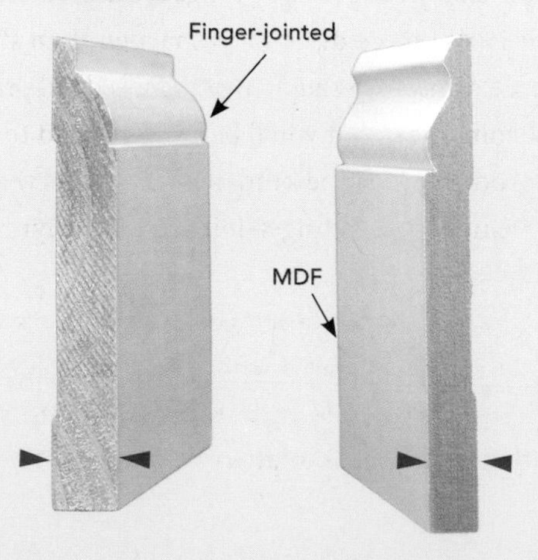

Sharp knives, well-balanced molding heads, and slow feed rates yield the smoothest, most-uniform molding. But maintaining these optimum manufacturing conditions is more costly, as you might expect. When molding shows up on the job with a rough, chattered surface, it's usually an indication that feed rates were too fast or that molding heads weren't properly balanced. Dull knives also can cause rough surface texture as well as profiles that aren't uniform.

It's important to be aware of these quality-control issues when you're buying finger-jointed trim. Trim carpenter and *Fine Homebuilding* author Gary Striegler always inspects a large trim delivery. "If I see chatter on a board, I return it for a new one," he says. "To take out even subtle waves in a piece requires sanding. That's time and labor for me."

MDF molding tends to have softer profiles

MDF is made by combining an adhesive resin with wood shavings, chips, sawdust, and other mill waste. The mixture is formed, compressed, and cured in long, flat sheets; then the sheets are ripped and shaped in a molding machine like finger-jointed trim. A Scientific Certification Systems (SCS) rating indicates an environmentally friendly product; this also applies to finger-jointed trim.

According to the folks at Pac Trim®, one of the largest MDF trim manufacturers in the country, MDF dulls steel molding knives quickly, so it is always shaped using carbide knives. Carbide knives are considerably more expensive than steel knives, but they keep an edge much longer. Unfortunately, carbide is also more difficult to sharpen than steel and more difficult to shape into sharp-edged profiles. The bottom line is that you'll be hard-pressed to find MDF products with the kind of sharp edges that are available in high-end finger-jointed and polyurethane options.

Carbide knives that don't need frequent sharpening and inexpensive material explain why MDF molding tends to be less expensive than finger-jointed or synthetic versions. If high-end finger-jointed trim isn't within your budget, MDF might be a good alternative. It's incredibly stable, is available in a wide variety of profiles, and looks surprisingly good.

Although he prefers using solid lumber for trim, Montana carpenter Chris Whalen has no qualms about using MDF. "The long lengths are a little floppy to work with, but it's inexpensive, easy to cut, and takes paint really well," Whalen says.

COMBINING COMPATIBLE PROFILES

Creating the right tone with moldings requires careful symmetry among base, casing, crown, and other profiles. The wrong combination can be jarring. Most manufacturers offer sets of compatible profiles, but they aren't always historically accurate or even pleasingly proportioned. Windsor Mill has filled this void by creating well-researched, historically correct collections in popular styles such as Craftsman (see the photos below), Greek revival, and classic colonial. Their finger-jointed trim takes the guesswork out of finding compatible profiles (www.windsorone.com).

Katz and Striegler said similar things, and both like MDF because of its stability. But Striegler doesn't like casing doors with MDF because he thinks it's too susceptible to damage in high-traffic areas. Katz feels differently. "The notion that MDF isn't all that durable is a myth. I have used it everywhere, except in bathrooms," Katz says. It's no myth that if MDF gets wet, it swells, bubbles, and falls apart.

Synthetic trim offers the most-ornate profiles

Unlike finger-jointed and MDF profiles, which are formed by cutting away the material with molding knives, synthetic trim is either extruded or poured into a form. The extrusion process is used for flat trim and simple profiles. Most of these moldings are PVC-based and can be used indoors and outdoors. Most polyurethane molding can be used outdoors, but it's more commonly used for interior applications. Unlike PVC material, polyurethane trim doesn't expand and contract significantly in response to temperature changes. This helps to minimize gaps between joints.

Polyurethane is often used for ornate, highly detailed moldings that would be extremely expensive to fabricate in wood. Each length is made in a large rubber or steel mold. Once poured, the liquid expands like polyurethane glue. The more the liquid expands, the tighter the cell structure and denser the material.

The risk of mismatched profiles is one reason why some people choose higher-quality polyurethane products. To ensure uniform profiles, Century Architectural Millwork creates molds out of steel rather than rubber. Rubber molds wear out, causing a profile to change. Folks like Century and ORAC Decor® guarantee that profiles will match.

Because the raw materials and manufacturing process are more costly, synthetic molding tends to be more expensive than finger-jointed or MDF products. But it's also not a fair comparison because most synthetic trim is more ornate.

FINGER-JOINTED TRIM IS REAL WOOD WITHOUT THE KNOTS

By using a combination of finger joints (for end-grain connections) and edge joints (for wider molding), manufacturers create long blanks from which molding is shaped. The straightest, smoothest, stablest products are made from trees of the same species and similar ages. Some manufacturers, like Windsor Mill, profile the bark side of a board, which they say minimizes the potential for grain to raise and show through a finished surface. Quality varies greatly with this type of trim. Watch out for visible joints, raised grain, rough texture, color variation, and uneven primer coverage. Many local or regional lumberyards mill their own finger-jointed trim, so it's not necessarily sold by brand name. Windsor Mill (www.windsorone.com) and Moulding & Millwork (www.mouldingandmillwork.com) are two major manufacturers that distribute nationally and maintain useful websites.

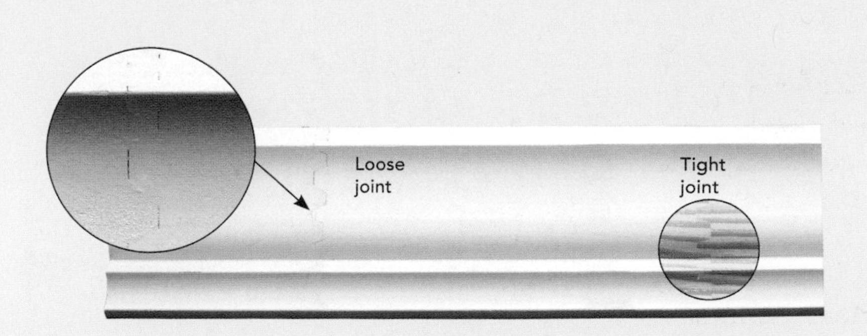

TIGHT JOINTS ARE ESSENTIAL FOR STRENGTH AND APPEARANCE. Products like this have given finger-jointed trim a bad name. Loose joints can be caused by poor wood selection, machining problems, or adhesive failure.

Edge
joints

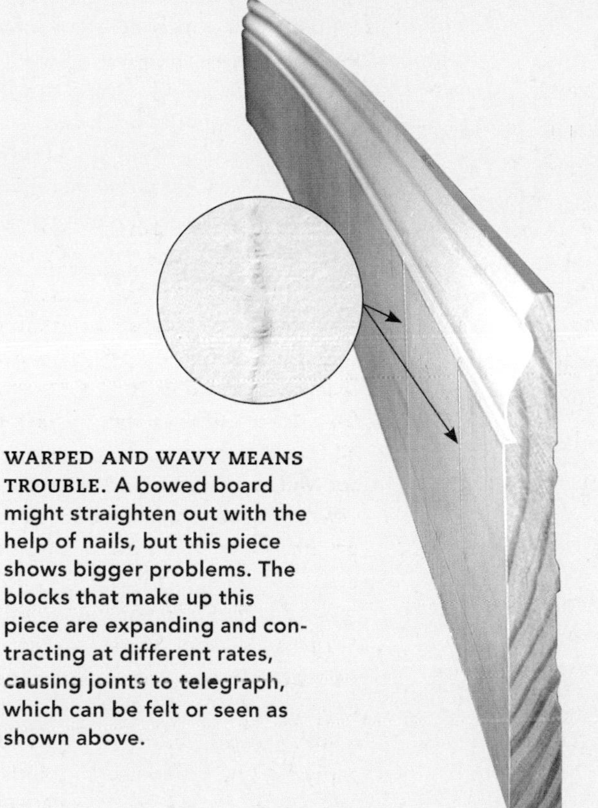

WARPED AND WAVY MEANS TROUBLE. A bowed board might straighten out with the help of nails, but this piece shows bigger problems. The blocks that make up this piece are expanding and contracting at different rates, causing joints to telegraph, which can be felt or seen as shown above.

FINGER JOINTS PROVIDE PLENTY OF GLUE AREA. Precisely cut fingers are the key to tight-fitting joints. Wide shoulders on adjoining pieces, indicated by the circled areas, strengthen the joint.

PROS

- Installs and handles like solid wood.
- Less prone to cupping and warping than real wood.
- Good nail-, screw-, and glue-holding properties.
- Can be used in moist locations.
- High impact resistance.
- Higher-end options tend to offer better-defined profiles than MDF.
- Readily available in 16-ft. lengths.

CONS

- Heavier than lower-density MDF and synthetic trim.
- Lower-end products tend to be less stable than MDF or synthetics.
- Finger joints can telegraph through the finish in lower-quality products.
- Wood grain can show through in lower-quality products.

MDF MOLDING IS SMOOTH AND STABLE

MDF IS MADE OF A WOOD-FIBER and resin mixture that's compressed into long, flat sheets. After the sheets are ripped into appropriate widths, they're run through a molding machine to form the profile. High-density MDF trim is a smart choice in high-traffic locations and for long, straight runs. Lower-density products are lighter weight, but they aren't as durable and don't hold fasteners as well. Many local and regional lumberyards purchase or mill their own MDF trim, but several manufacturers, like Burton Mouldings (www.burtonmoldings.com), are distributed nationally. Although strictly a wholesaler, Pac Trim (www.pactrim.com) has some useful installation information on its website. Some companies, like White River™ Hardwoods (www.whiteriver.com), offer MDF crown in ornate profiles with embossed detailing.

COPING WITH NAILING. MDF's uniform composition makes cope cuts easy. However, sharp corners and curves are prone to damage as the joint goes together. Pneumatic nails can cause mushrooming or surface damage in lower-density MDF.

MANUFACTURED CORNERS TAKE AWAY THE GUESSWORK. Some manufacturers offer systems for turning corners or joining two lengths of trim using blocks. Shown here: PacFit Components.

PROS

- High-density MDF copes and cuts much like wood.
- Can be joined with traditional wood glues.
- Can be filled or patched with the same fillers used for solid wood.
- Extremely stable; won't cup, warp, or shrink much.
- Smoother primed surface than on most finger-jointed trim.
- Readily available in 16-ft. lengths.

CONS

- Not suitable for moist areas like bathrooms.
- More cumbersome to work than finger-jointed or synthetic trim because lengths can flop around easily.
- MDF dust is even more of a nuisance than wood dust.
- Pneumatic fasteners can mushroom or chip away the primer coat.
- Small pieces, such as mitered returns, are even more fragile than wood.
- Pilot holes are necessary if hand-nailing.
- Profiles lack crisp edges.

Former trim carpenter turned luxury-home builder Peter Ziamandanis has been using polyurethane moldings for years. He likes poly products because large built-up cornice details are available as one piece, saving labor and often material costs. Ziamandanis says that installing polyurethane moldings is as easy as installing wood or MDF products. "But I always miter inside the corners. You can cope polyurethane much like wood, but it's difficult to do with highly detailed trim. Because poly moldings are incredibly lightweight, you don't need to hit a stud to install them. Adhesive caulk and a couple of crisscrossed nails will hold it just fine," he says.

Because polyurethane-based trim tends to be soft and is damaged easily, many people consider it best used in out-of-reach areas, such as cornices. But some manufacturers (ORAC Decor for one) have high-density products durable enough to be used for casing and baseboard.

Thick, even primer coats should be the standard

Manufacturers of finger-jointed and MDF products prime their molding with a similar process. Once the profile has been milled, the newly shaped boards ride a conveyor belt through a box where paint is applied; then they go through a large drying oven.

MDF generally isn't back-primed, but because it's such a stable material, it won't cup or warp, so back-priming is unnecessary. Unlike finger-jointed products, though, MDF is sanded and buffed between coats because cutting profiles raises the fibers, which swell once they're moistened with paint. You're unlikely ever to see a raised, rough edge on a piece of MDF trim. But you might find pieces that haven't been coated evenly.

Molded polyurethane products have a bonded primer coat. The mold is sprayed with the primer before the liquefied poly is added. Because of this process, primer won't likely chip from the surface. The primer also adds to the durability of the product.

SYNTHETICS ARE LIGHTWEIGHT AND IMMUNE TO MOISTURE

Unlike finger-jointed and MDF molding, synthetic trim is either extruded or molded. PVC and polystyrene molding is usually extruded to produce flat trim and simple profiles. It can be stamped to add detail, but profiles are usually not as clean and as crisp as what you'd find with polyurethane. Look for products that guarantee that factory-cut ends will match up perfectly at butt joints. For high-traffic areas, look for high-density products recommended for such uses, such as PVC or polystyrene. Some products meet code-required fire-spread ratings, or can be coated so that they do. Major manufacturers include Century™ Architectural Specialties (www.centuryarchitecturalspecialties.com), Fypon® (www.fypon.com), and Architectural Products by Outwater® (www.outwater.com).

FLEX MOLDING ELIMINATES THE NEED FOR KERFS

When running trim around curved walls or arched openings, check with the manufacturer of the product you're planning to use. They likely offer a curved option for most profiles. Made to match finger-jointed, MDF, and synthetic profiles, flexible pieces are installed the same way as their rigid cousins. But don't count on the profile to line up perfectly. Joining flex molding to a rigid version requires sanding and filling. Order enough to give yourself room for error.

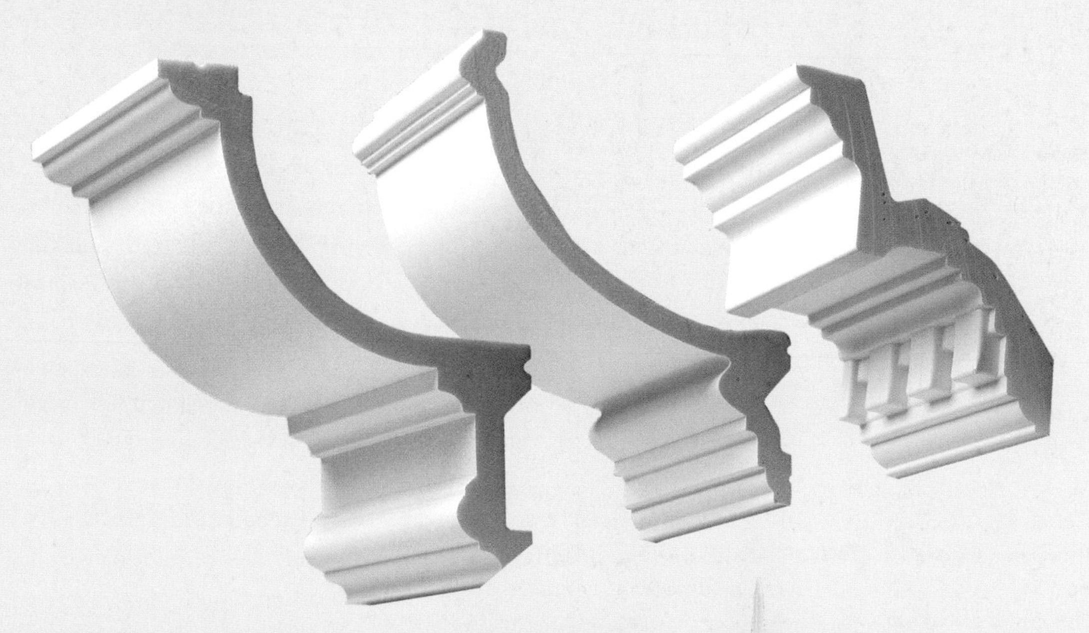

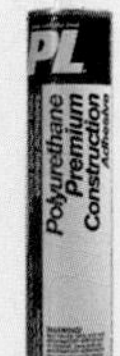

BUILT-UP CROWN IN ONE LIGHTWEIGHT PIECE. Poured into molds, liquid polyurethane expands and hardens to fill out intricately detailed profiles that would be expensive and time-consuming to create in wood. Manufacturers recommend installing and joining polyurethane molding with expensive proprietary adhesives, but a $5 tube of PL® Premium Polyurethane Construction Adhesive (www.stickwithpl.com) works just as well.

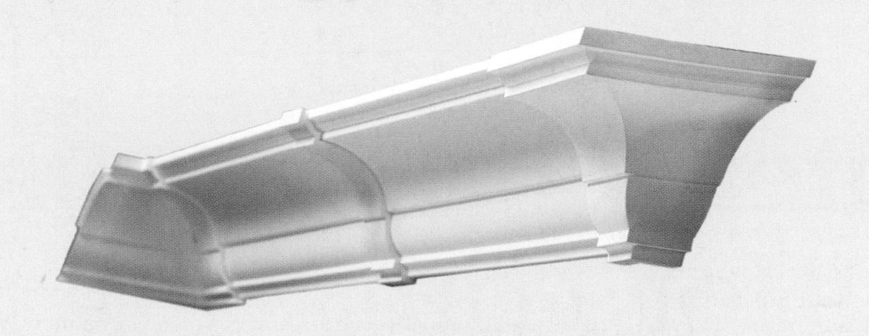

ORNATE PROFILES CAN BE TOUGH TO JOIN. Many manufacturers offer formed corner and butt-joint pieces to simplify installation. Shown above, an interlocking crown-molding system in the ORAC Surroundings series (www.outwater.com).

PATTERN REPEATS SHOULD ALIGN. When they don't, filling and sanding are required. All of Century's moldings (including the egg-and-dart crown shown above) are guaranteed to align when butt joints are made as recommended.

PROS

- Cuts and nails like wood.
- Impervious to moisture; most types also can be used outdoors.
- Lightweight; some types can be installed with adhesive only.
- Polyurethane trim is more dimensionally stable than other types.
- Will not warp, crack, or splinter.
- Comes in the most-ornate profiles.
- Limited warranty available for most products.

CONS

- More difficult to cope than MDF or finger- jointed trim.
- Expensive proprietary adhesive sometimes required.
- Not available in lengths longer than 14 ft.; some profiles are available only in 6-ft. lengths.
- Polyurethane is less durable than finger-jointed or MDF trim.
- Repairs and nail holes often require special filler.

4 Router Tricks for Trim

BY GARY STRIEGLER

My editor recently called me a tool junkie. My wife has called me the same thing, but with a few expletives. In my defense, when you've been building as long as I have, you can appreciate how important it is to have a good collection of tools close at hand wherever you are.

I won't tell you how many routers I actually own, partly because I stopped counting 10 years ago. But despite the grief routers might cause me on occasion, they have gotten me out of a lot of jams on the job. Of the many I own, the trim and midsize D-handle models are the two I reach for most.

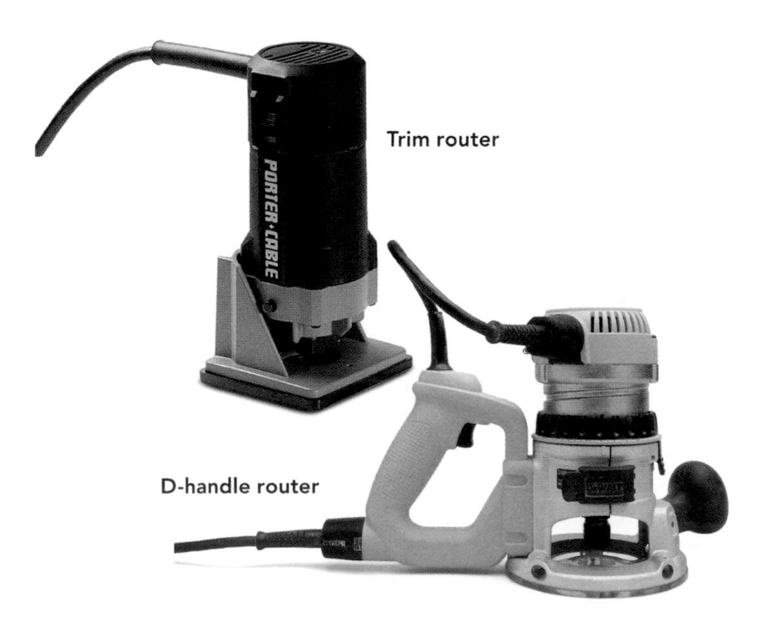

Different routers for different jobs

The first trim routers were called laminate trimmers because they made quick, clean work of flush-cutting laminate for countertops. Today, more powerful motors and a variety of accessories make trim routers a staple for most finish carpenters. One of my favorite features is that I can safely use my trim router with one hand, leaving the other hand free to hold the workpiece.

I use a trim router when I need to make short, shallow cuts in doors, cabinets, and trim. Because most have only 1-hp motors and ¼-in. collets, they shouldn't be commissioned for heavy work like cutting detailed profiles into a piece of hardwood. I don't use anything larger than a 1-in.-long by ½-in.-dia. straight bit in my trim routers. Larger-diameter bits put too much strain on the motor.

For bigger tasks that require more power, such as cutting deep mortises or plowing wide dadoes, I use a 1½-hp D-handle router. The D-handle allows me to use one hand or two, and it is easy to control.

If you don't own either of these tools or the bits I mention here, show this article to whoever is likely to give you a hard time about a tool purchase. Together, they'll save you an incredible amount of time, which will let you find reasons to buy more tools.

1. Patch blowouts on trim after it's installed

Pneumatic nail guns are nice, except when fixing an errant nail adds to my list of things to do. Bondo® or wood filler works when the blowout is not large and deep. But nothing does the trick quite like a dutchman, or wooden patch, especially when the blowout happens in the jamb reveal (shown here). I used to use a hammer and chisel to make the mortise for a dutchman. Now, I plow a better mortise in half the time with a trim router equipped with a ½-in.-dia. pattern bit. The bit doesn't need to have a large cutting length; I usually use one about ½ in. long to keep the router closer to the cut, which gives me more control.

When working with stain-grade trim, I make the dutchman from the same material and match the grain. If it's paint-grade trim I'm patching, I don't worry about wood type too much.

I often use headless pins to provide extra holding power for the dutchman, but I've seen them clamped in place with masking tape as well. Either way, after the glue dries, a quick pass with 120-grit paper on a random-orbit sander makes the patch practically disappear.

This approach also works for replacing damaged sections of profiled trim, especially if the trim is already installed.

THE BITS I USE THE MOST

Bits equipped with a bearing that matches the cutting diameter follow a template aligned to the cutline, which makes setup quick and easy. **Pattern bits** have a bearing mounted on top of the cutting area that follows a template mounted to the top of the workpiece. **Flush-trim bits** have a bearing mounted on the bottom that follows a template mounted below the workpiece. **Dado bits** aren't equipped with a bearing, so they require the use of a guide for the router. Their deep, nearly hollow center allows waste to exit the dado as you cut. No matter which style you're using, buy carbide-tipped bits that have two flutes. Carbide lasts longer, and two flutes make a cleaner cut than one.

Pattern

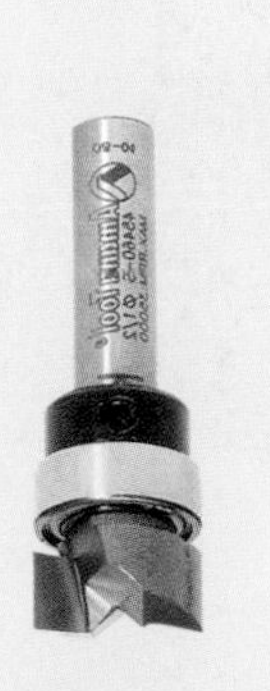

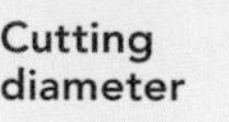

Cutting diameter

Flush trim

Cutting length

Dado

Flute

SET THE DEPTH. Nail an MDF guide strip to the jamb using headless pins; then adjust the bit depth to match the jamb's reveal.

PLOW A MORTISE. With the router on, push the bit into the jamb until the bearing registers. Then follow the guide until the blowout is gone.

FIT THE DUTCHMAN. It's faster to radius the dutchman than to square the corners of the mortise. Make the dutchman a bit thicker than the mortise's depth.

FASTEN AND SAND. Glue and nail the dutchman with headless pins. Once the glue dries, sand the patch flush with the jamb.

2. Get clean, evenly spaced dadoes to make dentil molding on site

I hang a lot of custom built-up crown that includes dentil molding, and I often make the dentil molding on site. Some carpenters do this by ripping multiple dadoes on a tablesaw, but I have found that using a router is faster, is more accurate, and produces cleaner cuts.

I plow dadoes across a length of 1×6 poplar; then I rip the board to width. To space the dadoes evenly, I attach a subbase and a guide strip to my D-handle router. The subbase creates a wide, stable surface for cutting, and the guide strip rides in the previously cut dado. Centering the subbase on the router ensures that at least 3 in. of the strip locks

USE A SUBBASE AND A GUIDE STRIP. With a square as a guide, start by plowing a starter dado on one end of the board, then make an oversize router base with a 12-in.-sq. piece of ¾-in. MDF. In the center of the MDF, drill a hole larger than the diameter of the bit you're using, and run screws through the router's base to secure it. Then attach a strip of clear stock, such as pine or poplar, that fits the first dado; it should fit snugly but slide freely. To make it easier to register the template in the dado, run the strip just past the subbase so that it's visible. A few headless pins hold it in place.

into the dado before the bit engages and as it exits the cut.

The size of the bit I use depends on the size of the dentil molding I'm making, but I prefer to use bits with a ½-in.-dia. shank; smaller shanks can flex, which affects the cut. They also break more easily when plowing large dadoes.

3. Trim a cabinet face frame in place

When rough openings for appliances are too small, they often need to be widened on site. Modifying cabinet face frames can be nerve-racking, especially when the cabinet is already installed. I've tried plenty of approaches to this task, many of which left me disappointed with the results. The most precise way I found is with a ½-in.-dia. flush-trim bit and a trim router.

I pin a piece of flat stock, typically MDF, to the cabinet behind the stile I'm cutting as a guide for the bearing. Keeping in mind that the bearing is the same size as the bit's cutting diameter, I select stock based on the amount of material I plan to take off (typically 1⁄16 in. to ⅛ in. at a time).

I like to use a trim router here because I can grip it with both hands for complete control while still having clear sight of the bit as it's removing material. If I need to take more than ⅛ in. off at a time, I switch to a D-handle; the added power produces a cleaner cut.

MOVE IN THE RIGHT DIRECTION. When the bit is to the left of the stile, as it is here, move from the top down. When it's to the right of the stile, move from the bottom up. Keep both hands on the router, and move slowly. Straight bits produce a radiused corner (see the photo at right). The radius size depends on the bit's diameter and the amount of material you're taking off. Use a ½-in.-dia. bit to keep the radius as small as possible. The smaller it is, the less work you'll have to do with a chisel.

4. Mortise tricky door hardware with simple jigs

It's easier to mortise a latch plate, ball catch, and slide bolt before hanging the door. When mortising a latch plate, I dismantle the jamb before it's installed and set it up on a workbench. A short pattern bit and a trim router are good for latch-plate and ball-catch mortises. Slide bolts require a deep mortise, so I use a longer pattern bit and a D-handle router.

MAKE A SELF-CLAMPING JIG. Cut a dado into a piece of ¾-in. MDF to register on the doorstop. Then make a cutout the same size as the latch plate. Hold the template down with one hand, and plow the mortise with the other. Use a corner chisel to square radiused corners with a quick tap of the hammer.

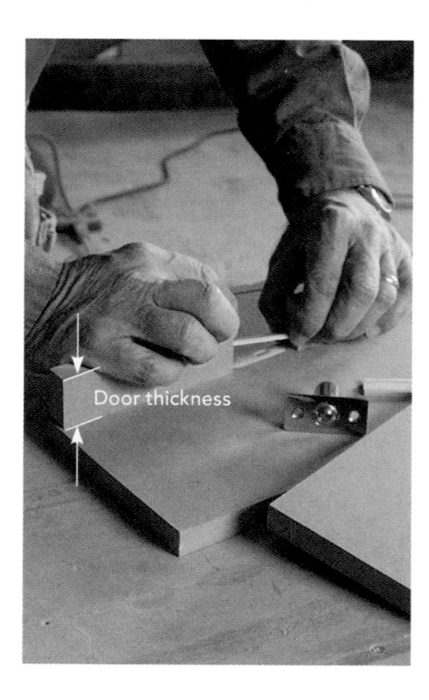

BALL CATCHES GET A SADDLELIKE THREE-PIECE JIG. Nail and glue two 4-in.-long pieces of ¾-in. MDF, ripped to match the door thickness, in between two pieces of ¾-in. MDF. Space the ripped sections apart the length of the ball latch. Glue shims inside the opening on each side piece so that the opening width matches the ball-catch plate. Center the jig on the hole, and clamp it in place. Place the router onto the jig, locating the bit in the hole. Make a test cut near the hole to double-check the depth.

HOUSE TREADS IN A SKIRTBOARD FOR A TIGHT JOINT

I BUILD STAIRS ON SITE by wrapping framed stringers with hardwood treads and plywood risers. One of the most challenging parts of this process is creating a tight fit between stair treads and a skirtboard. You can scribe and butt the tread to the skirtboard, but it takes a lot of time and doesn't account for seasonal movement of the stringers or treads, which can cause the joint to open. For a better joint, I house the treads in the skirtboard. [1] After scribing the skirtboard to the stringers, I plow a dado for each tread using a template and a 1⅛-in. pattern bit. The bit creates a radiused end for the tread's bullnose, so I make sure to stop the bit exactly where the tread needs to end. The guide should be rigid enough that it won't flex and has to be perfectly flat, so I make one by gluing up a couple of pieces of 6-in.-wide ¾-in. MDF. [2] After installing the skirtboard, I test-fit each tread. I shim the tread as needed to push it tight to the top and front of the dado.

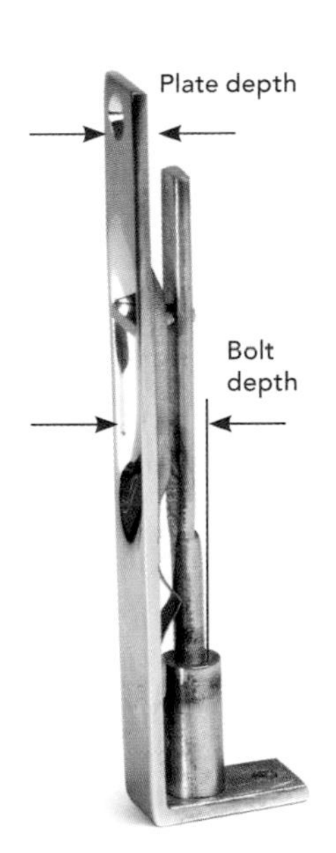

SLIDE BOLTS GET A SIMILAR JIG AND TWO CUTS. Slide bolts like the one shown here sit in a mortise cut into the door's corner. Use a pattern bit with a cutting diameter that matches the width of the slide-bolt plate. Make the first cut with the bit set to the plate depth.

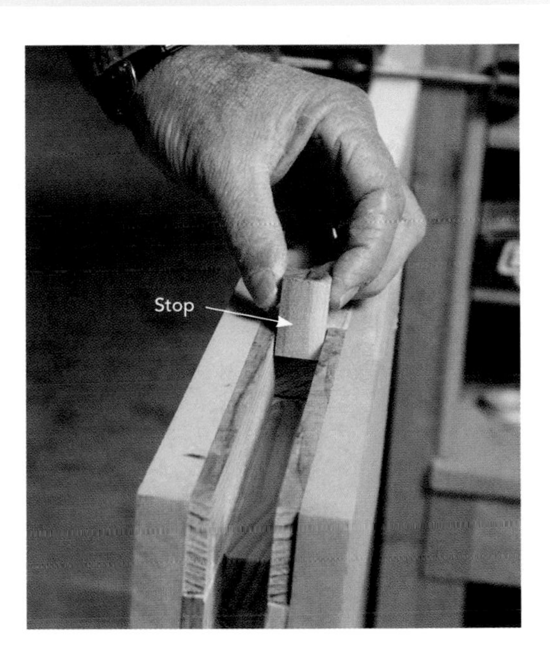

ADD A STOP FOR THE SECOND CUT. Make a second, deeper cut to the bolt's depth. Before making the cut, drop a stop into the jig to keep the bit from carrying through the end of the shallow cut where the plate will register. Make this cut in two passes.

Signature Trim Details

BY CHARLES BICKFORD

When you build a house, the interior trim is one of the last things to be installed, and one of the first things you notice after it's complete. Because it's always visible, interior trim deserves a fair amount of effort in both design and execution. For the designers who drew the plans and the carpenters who installed it, these details showcase their craft to an audience every day, a calling card of sorts. For those chronically afflicted with SRS (serial remodeler's syndrome), a new trim scheme is a great excuse to set up the miter saw.

As building science matures, it's getting harder and harder to build a house, let alone build one that has personality and a unique sense of character. But of all the materials used to build a house, the trim can be one of the least expensive and the easiest to customize.

Here, we're featuring a gallery of ideas that show how some designers and builders are stretching the traditions of trim.

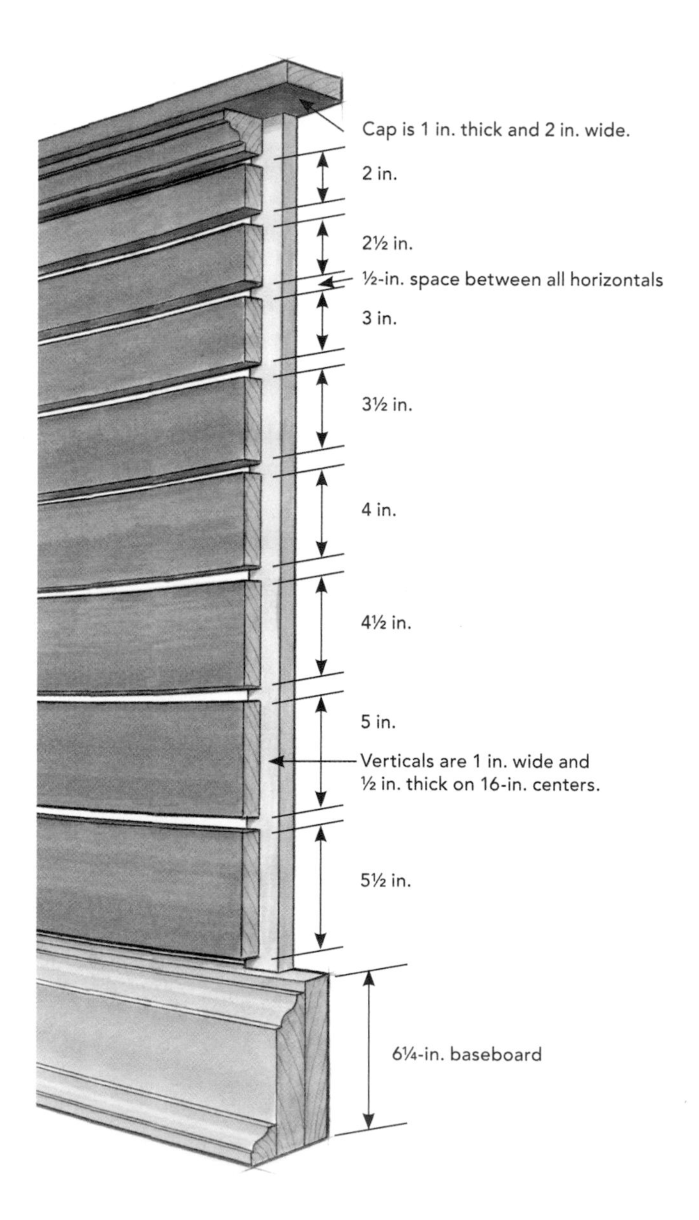

Design: Shope Reno Wharton Architecture, South Norwalk, Conn.
Construction: Newport Housewrights, Middletown, R.I.

1. Reinterpreting tradition

Bernard Wharton planned his house on Narragansett Bay with both a sense of tradition and a desire to experiment. He says, "You want a sense of timelessness…a strength through simplicity." Taking his cue from both boat-building and Shaker furniture, Wharton drew up plans for the wainscot in the entry hall. Like a Shaker chest of drawers, the horizontal boards get progressively narrower and give the wainscot a certain lift. The overall lightness and depth is increased by the airspace between the wall and the mahogany boards.

2. A recess where trim meets furniture

Built-ins walk a fine line between furniture and trim; they should both blend in and stand out. Here, a dining-room sideboard was built into a recess framed by a two-part molding. Instead of applying the molding over the drywall, however, the designer first placed flat drywall pilasters that were recessed back from the main wall plane. He then added the molding along the perimeter that's just proud of the wall surface. At the head, it folds in and becomes a valance across the opening.

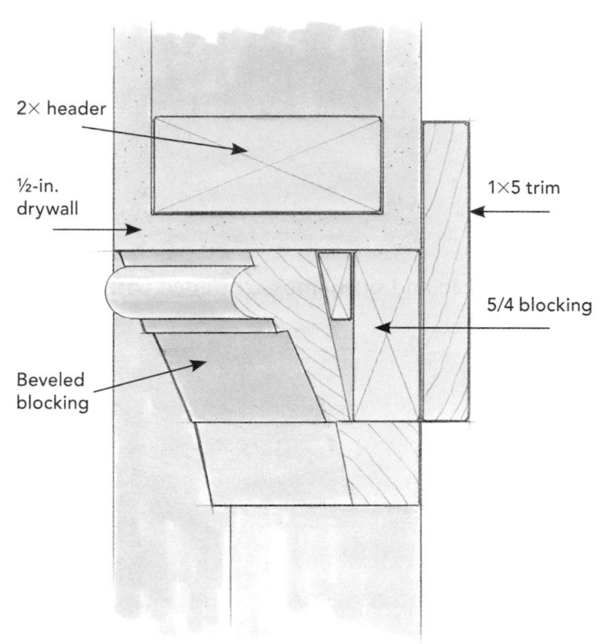

Design: Daniel Colbert, Riverside, Conn.
Construction: David O'Sullivan, Fairfield, Conn.
Sideboard: Breakfast Woodworks, Guilford, Conn.

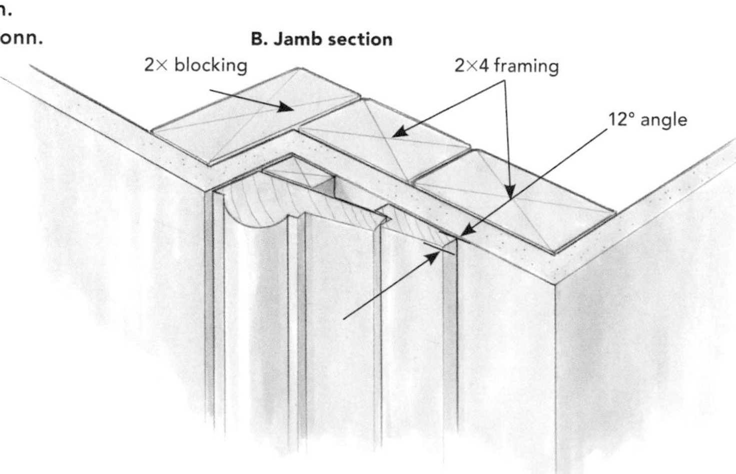

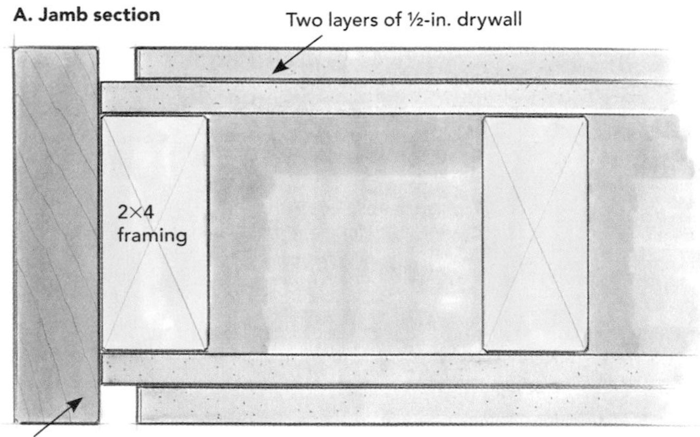

Design: Ennead Architects, New York
Construction: Doyle Construction, Vineyard Haven, Mass.

3. Straight and narrow

Modern design often reveals a different path to interior trim, relying on structural elements instead of ornate moldings. Richard Olcott of Ennead Architects (formerly Polshek Partners) designed this house around the concepts of clean lines and space. Where casing would cover the seam between jamb and drywall, a ½-in. reveal creates a clean shadowline. The head and side jambs are joined with a simple rabbet.

4. Looking to the East

On a trip to Japan, Scott Simons toured centuries-old tea-houses and temples, and was so taken by the designs that he sketched details and faxed them back to the office as he went. It wasn't long before he had a project where he could marry a modern design with a Japanese aesthetic. Here, plywood panels are used as a Western take on the traditional Japanese shoji screen. At the floor, maple baseboards are scribed into round mahogany columns that punctuate the flow of this hallway. A flat horizontal cap accentuates the base.

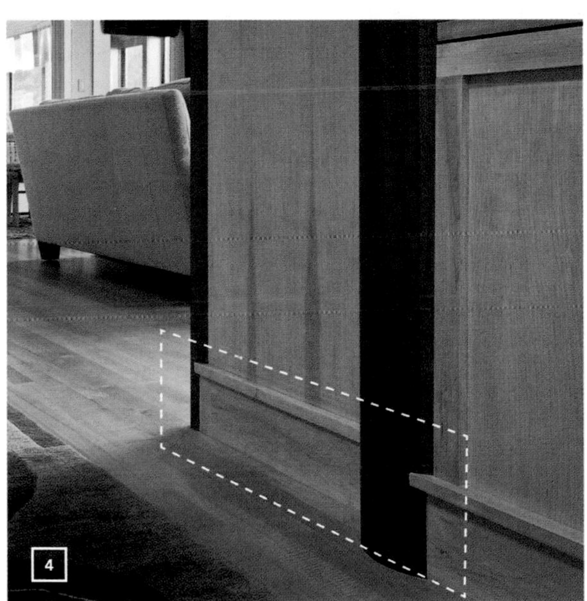

Design: Scott Simons Architects, Portland, Maine
Construction: Rousseau Builders, Pownal, Maine

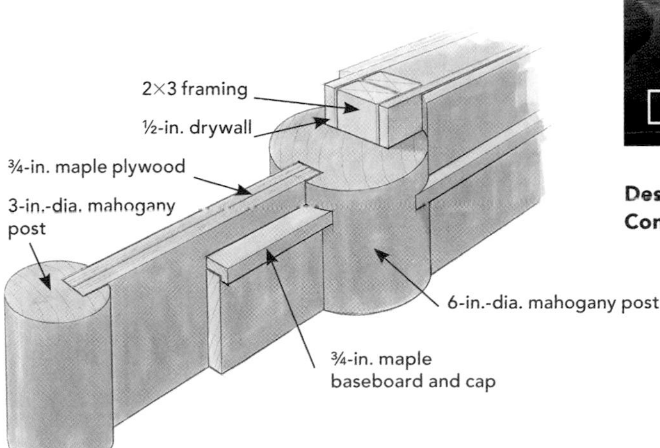

5. When less is more

Sometimes suggesting an interior-trim detail is enough to get the idea across. Minimal ornamentation often reveals the lines of the structure itself, and the geometry of the building becomes its own decoration. Around these skinny windows, the drywall butts directly to the window frame at the sides of the jambs. Only the sill and head are cased. This treatment may be preferred for its clean look, but it also makes sense in situations where curtains or drapes would cover up whatever decorative trim had been applied.

Design: Daniel Colbert, Riverside, Conn.
Construction: David O'Sullivan, Fairfield, Conn.

A. Head section

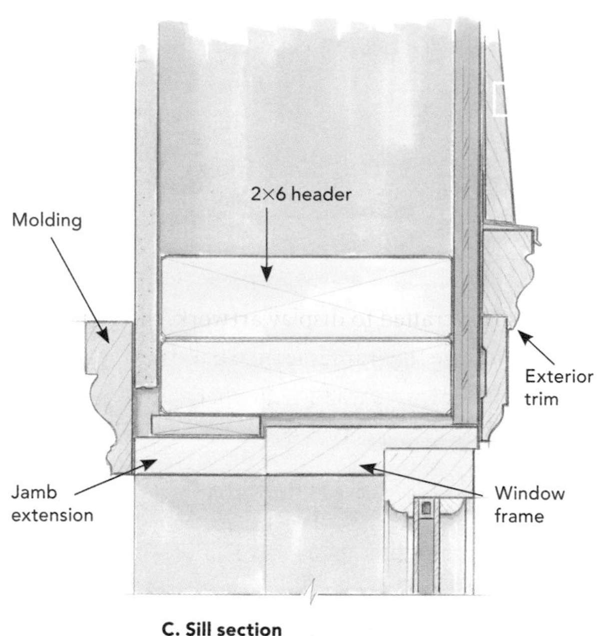

C. Sill section

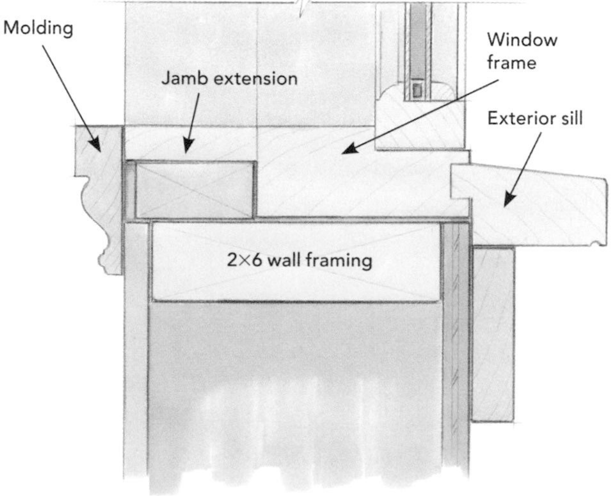

B. Jamb section

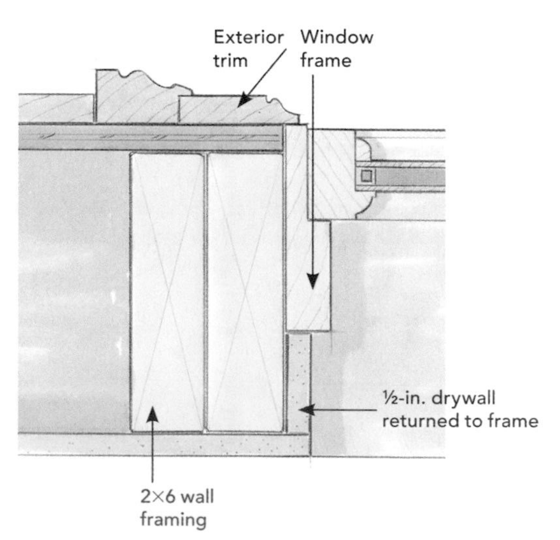

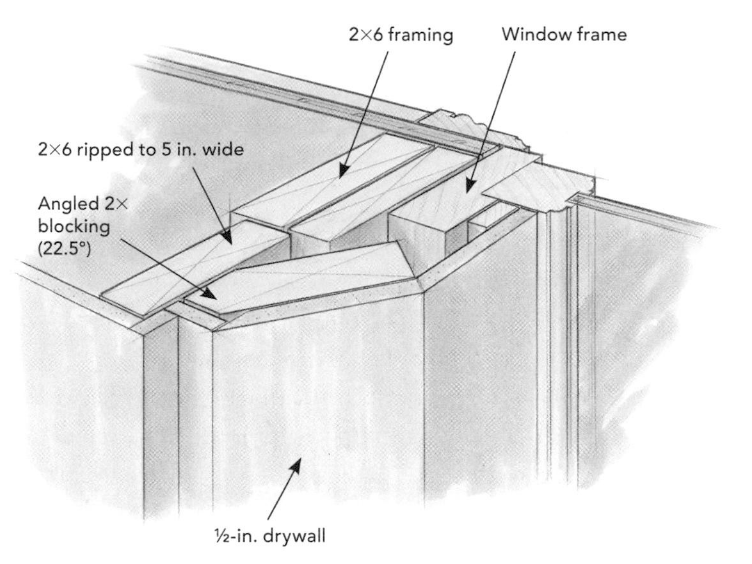

6. Frame the art, not the window

Windows are usually picture-framed with trim, but in this space crafted to display artwork, the designer decided to save the frames for the paintings. Architect Daniel Colbert says that the shadowline makes the windows recede, while the thicker wall becomes "a more substantial ground for the art."

This detail is accentuated by a narrow fold in the drywall that creates a shadowline. A continuous line of molding carries across the window tops, visually connecting each window to the hallway.

Design: Daniel Colbert, Riverside, Conn.
Construction: David O'Sullivan, Fairfield, Conn.

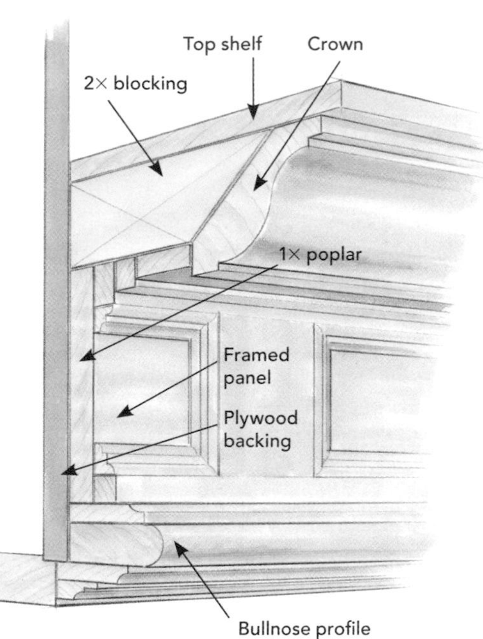

Design: Leon Trice, BT Architecture, Portland, Ore.
Mantel: Mark Newman, Design in Wood, Portland, Ore.

7. The contrast of a single curve

Houses are full of trim. There are windows, doors, and baseboard in just about every room. Overdoing these details can make a house busy and uncomfortable. On the other hand, a fireplace mantel is a great place to have some fun. An eye-level detail, a mantel deserves and receives a level of attention enjoyed by few other elements.

While designing a guest house on the Willamette River near Portland, Ore., Leon Trice thought that "things get repeated too much" in general and didn't always want to rely on geometry. Instead of a stock profile below the mantel, he hand-drew an irregular curve that echoes the crown above yet has a graceful flow all its own. The curve's sweep is accentuated by slightly thicker stock. He used a similar design on another mantel (see top right photo) on the same project.

PART 2

Baseboard and Crown Molding

Installing Baseboard

BY JOHN SPIER

Most trim carpenters I know started their trade by learning to run baseboard, and the first baseboard they ran was probably in a closet. You might think this happened because baseboard is easy to install or because mistakes are harder to see near the floor, especially in a closet. The real reason is that the teacher has bad knees and probably doesn't fit so well in the closet anymore.

There are lots of types of baseboard, from a simple 1×4 to elaborate assemblies of multiple moldings. To illustrate the basics, the baseboard featured in this article has a flat profile with a simple bead along the top, and I installed it in a new house.

Which comes first: Flooring or baseboard?

Whether flooring or baseboard comes first depends on the type of flooring in each room. For wood floors, I install the floor first so that I can sand the floor edges and then run the base. I also install and grout tile before the baseboard goes in, unless the tile is so rough and irregular that I would have to scribe the bottom of the base to fit the tile. In this case, I run the base first, finish it, and then tile and grout to it.

In areas prone to wetness, tile should be installed first so that water doesn't collect around the bottom of the baseboard. Because vinyl flooring is so easily damaged, I don't like working over it. In rooms that will receive vinyl floors, I often cut and fit pieces of baseboard and then take them away to be painted while the vinyl installer works. Carpeted floors are generally the easiest. I install and finish the baseboard first, usually with a ¼-in. space underneath where the carpet layer can tuck the edges of the carpet.

Baseboard butts to door casings

In most simple trim jobs, baseboard butts to the outside of the door casings, so the casings must be installed first. Other things that baseboard runs into, such as built-in shelving, window seats, and fireplace surrounds, also should be installed or at least precisely located.

Stair skirtboards often tie into the baseboard and should be in place as well. When the trim is to butt into baseboard heating units, the plumber or electrician gives me the exact locations and dimensions for the spaces to be left out of the baseboard. Last, I check for any steps, ramps, landings, or other changes in floor level or material, such as where a carpeted living room makes a transition to a vinyl kitchen floor. If the difference is minimal, the baseboard can be ripped down to keep the top all at the same level.

KNIFE MARKS THE SPOT. Wherever possible, the author places a slightly oversize piece of stock in place and then marks the exact length with the blade of a razor knife.

Mark the lengths with a knife

I like to work from a pile of baseboard stock at a miter-saw table centrally located in the house. Here, I can keep offcuts organized so that I use the stock more efficiently. Almost invariably, I work my way around each room from left to right. I choose this direction because as a right-handed carpenter, I find that most base profiles are easier to cope on the left end of each piece.

My first piece starts at the right-hand side of a door casing and runs to a corner or to the next casing or wall interruption. There are a couple of ways to get the right length of each piece. The most obvious way is getting the exact measurement of each piece with a tape measure. Wherever possible, though, I prefer to take a piece of stock that's slightly long, put it in place, and then mark the exact length with a razor knife (see the photo above).

Coping is easier than it looks

Installing baseboard requires the same joints as the rest of the trim in the house. Inside corners are coped, outside corners are mitered, and long runs are joined with scarf joints. Simple butt joints suffice where baseboard meets stair skirtboards or casings. I often use a biscuit to align and strengthen a butt joint.

Coped joints are easier to make than they look, and they're actually quite forgiving. A coped joint has the baseboard on one wall running square into the corner with the adjacent piece scribed or shaped to fit into it.

The coped piece starts with a 45° bevel on the end that joins with the square-end piece, as if it were to be mitered for an inside corner. Then the excess beyond the mitered edge is removed so that the two pieces fit together.

I make the initial 45° cut with a miter saw (see the top photo on p. 45). Next, I flip the board on edge and plunge the miter saw down the straight part of

Mark both pieces.

Test the fit.

Long piece goes in first.

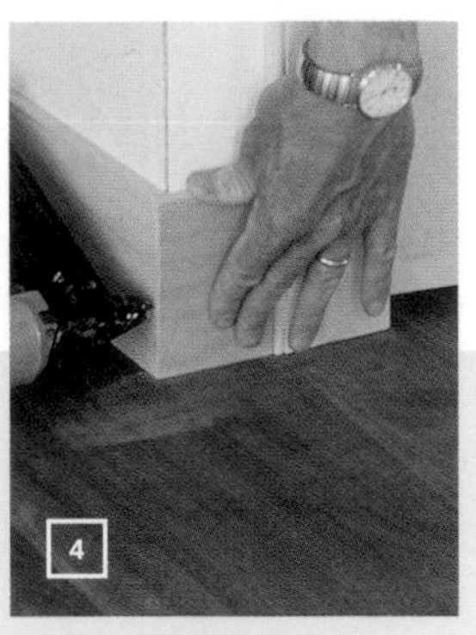

Glue and pin the joint.

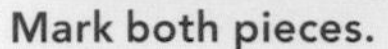

MITER OUTSIDE CORNERS

The quickest way to a tight outside baseboard corner is to scribe both pieces to the corner [1]. The author adjusts the miter cut by eye to make up for any out-of-square condition and follows the scribe lines to compensate for any differences in plumb. The pieces then are dry-fit [2], and the first piece goes into place [3]. The joint then is glued and pinned before the second piece is nailed to the wall [4].

the profile with the blade set at about a 5° back bevel (see the middle photo on the facing page). The excess for any decorative detail then can be cut back with a coping saw (see the bottom photo on the facing page). The bead on this baseboard was tough to cut perfectly with the coping saw, but I used a sharp chisel to clean it up. Again, only the top edge of the profile needs to be square and precise. The back bevel along the rest of the profile makes it easier to fine-tune the fit.

Working around the room means that only one end of each piece is coped. So after the cope is cut, the length is measured, and the other end is cut square. If that end goes to a corner, it is in turn covered by the next cope.

Don't measure for outside corners

Outside corners are mitered together, but I never measure these pieces. Instead, I cut them a few inches long, hold them in place, and scribe a line on the back of the piece for the cutline (see photo 1 above). I mark both pieces before fastening either one.

Often, corners are not exactly square, and baseboard isn't always exactly plumb. I usually can estimate how much I need to adjust the miter, and following the scribed cutline takes care of the out-of-plumb condition. For problem areas, I test the angles with two scrap pieces.

When I've cut the two outside corner pieces, I try them to make sure they fit and form a tight corner (see photo 2 above). Then I fasten the first piece in

COPE INSIDE CORNERS

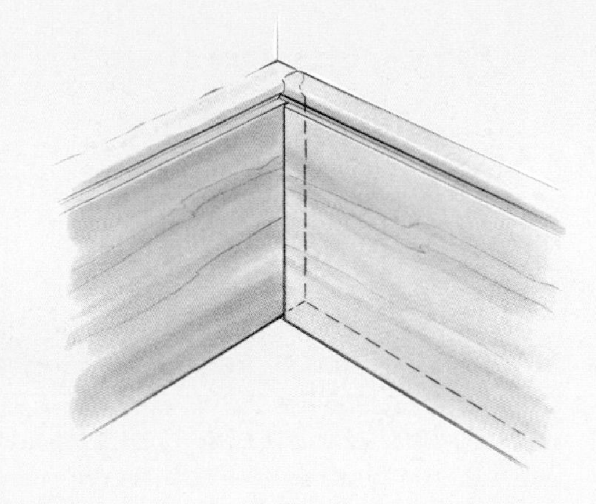

WITH THE FIRST PIECE CUT SQUARE and installed in the corner, the author cuts a 45° angle on the adjoining piece as if for an inside miter [1]. The piece then is flipped up, and the miter saw plunges down the straight part of the profile with a slight back bevel [2]. A coping saw then cuts the detailed part of the profile [3], and a sharp chisel or a rasp can be used to clean up small areas of the cut.

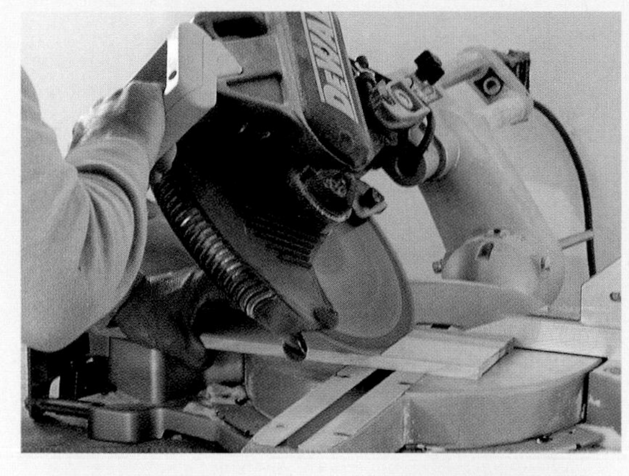

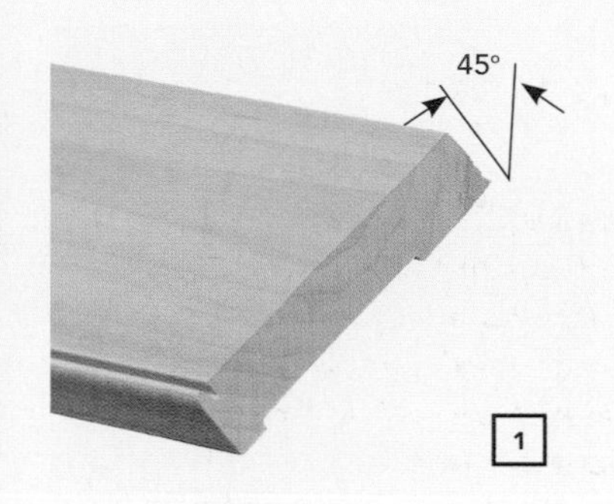

Start with a bevel. Using a miter saw, the author cuts a 45° bevel. This cut exposes the profile for the cope.

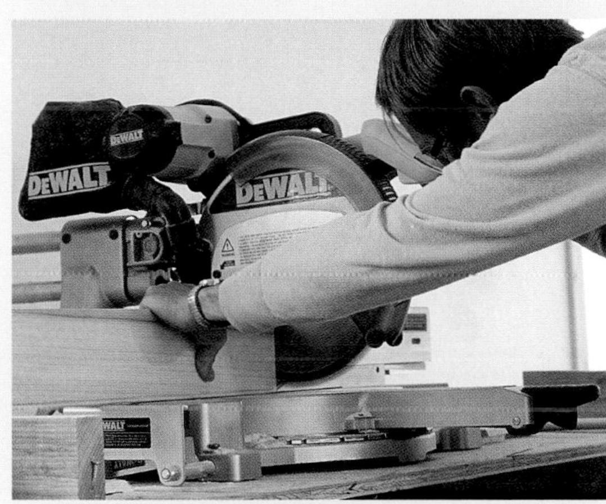

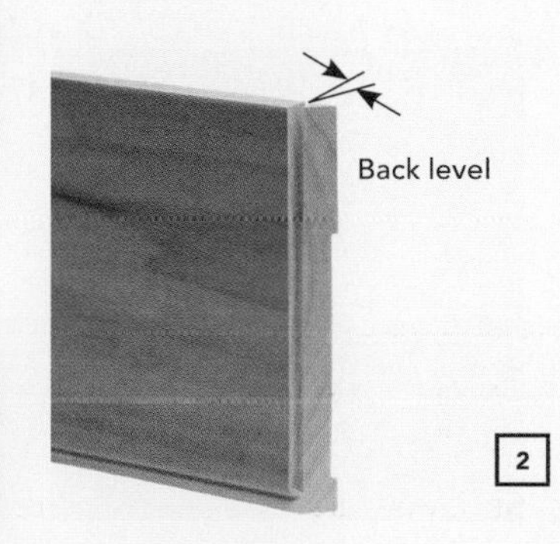

A flip and a plunge. The baseboard then is flipped upside down on edge, and the saw is plunged down the straight part of the profile at a slight back bevel.

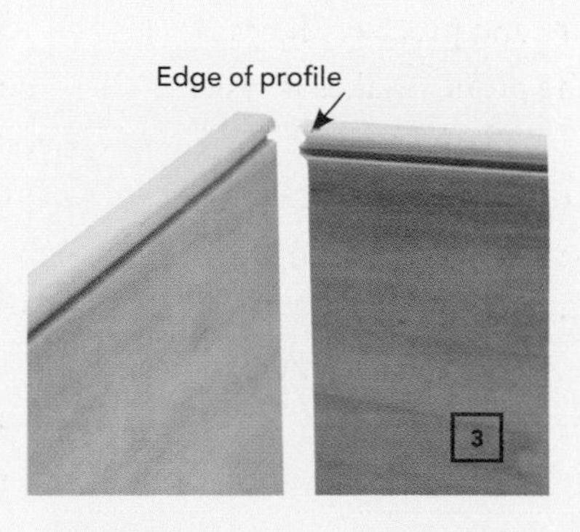

Cutting the profile. Following the edge of the exposed profile at a slight back bevel, the author uses a coping saw to finish cutting the profile.

DEALING WITH LITTLE PIECES

A little bevel is a big help. Putting a slight back bevel on a small piece of baseboard compensates for any irregularities in the corner so that it slips easily into place.

Narrow pieces shouldn't be nailed. Most small pieces that fit into corners cannot be nailed without danger of splitting, but the adjacent piece often can hold it in place without nails.

place (see photo 3 on p. 44). After running a bead of carpenter's glue along the joint, I install the second piece, pinning the corner together with a brad nailer (see photo 4 on p. 44).

Join long runs with a scarf joint

When a wall is too long for a single piece of baseboard stock, I break the run on a stud with a scarf joint, preferably in a place where the joint will be hidden—by furniture, for instance. A scarf joint is simply two pieces cut with 45° bevels so that the one piece overlaps the other. Scarf joints don't tend to separate like butt joints, and they look smoother.

When a scarf joint is called for and the run begins in a corner, I start with the coped end first. I make the 45° cut so that the face is the short side of the angle, and I tack that piece in place. Next, I line up a piece a few inches too long, butting the square end into the opposite corner, and I mark the distance to the backside of the cut, or the short side of the 45° angle (see photo 1 on the facing page). I test-fit the joint, adjusting the length on the square end to get the scarf just right (see photo 2 on the facing page). When I'm satisfied with the fit, I tack the pieces in

USE SCARF JOINTS ON LONG RUNS

When a run is too long for a single piece of baseboard, a scarf joint can join two or more pieces. The first piece goes in with a 45° angle cut into the end. The next piece extends past the first, and the length is marked [1]. After that piece is cut and dry-fit (see the far right photo), glue and brads secure the overlapping joint [2].

place, and then I glue and pin the joint with a brad nailer before nailing off the rest of the piece. If you try to glue that slippery glued bevel joint first, it's bound to slip out of alignment.

Another baseboard detail found in many houses is an intersection with a stair skirtboard at the top of a stair. Before it's installed, I cut the angle of intersection in the skirtboard. To keep the joint simple, I join the baseboard to the skirtboard with a butt joint, and I cut a small piece to test the fit (see the top right photo on p. 48). I then measure, cut the piece, and glue the joint before nailing it in place (see the middle right photo on p. 48).

Little pieces need a little help

I was taught to install baseboard using a hammer, a nail set, and 8d finish nails, but that was many years and two good knees ago. Now I fasten most of my base using a 15-ga. angled finish nail gun with 2-in. or 2½-in. nails. Along with a dab of wood glue, I fasten scarf joints and miters together with an 18-ga. brad nailer. Coped joints are supposed to allow a bit of movement, so they should not be glued.

Spaces next to door casings in a corner of a room require pieces of baseboard too small to nail. Also, these tight spots are usually out of square, so I cut the piece with a slight back bevel where it fits against the adjacent wall (see the top photo on the facing page). The adjoining piece of baseboard is usually enough to hold the short piece in place (see the bottom photo on the facing page), but I also use a shot of construction adhesive on the back of the piece for added strength.

Find studs before firing the nail gun

Before fastening baseboard, you should have a good idea where the studs are. There are lots of clues and tricks to locating studs. If I can still see the subfloor, I usually have the marks I made before the drywall was installed. If wood flooring has been installed, fastened to the joists, the studs are directly above those joists, at least on the bearing walls. I also frame every standard closet with studs 12 in. from the corners for the shelf cleats and one in the center for the pole bracket. In many cases, I'm able to take a common stud layout for the entire house from certain outside walls.

MARK LENGTH. 1

CUT TO LENGTH, TEST. 2

INSTALL. 3

But sometimes all these tricks fail, and I need to go hunting for the studs. On longer walls, I try to find one stud and then lay a tape measure on the floor next to the base as a guide to locate the rest. To find that first stud, I look for subtle clues in the drywall or plaster job, such as a seam or a dimple in the finish over a screw. Failing that, I look for an electrical box that's probably on a common stud (see the bottom right photo). A rap on the wall with my knuckle or a look inside the box lets me know which side the stud is on.

Before the base goes in, a nail driven through the drywall in a series of holes can locate both sides of a stud (see the bottom left photo). If you're looking for a stud after the base is in place, tapping a knuckle on the wall usually can get you close, and on a paint-grade job, I sometimes shoot an extra nail or two into the baseboard to help find the first stud. A quick-and-dirty trick I used to use on tract jobs was to tip the baseboard forward, rest the nose of the gun on top of it, and machine-gun a row of nails into the drywall behind the base. The first nail that didn't go clean through hit a stud.

BASE MEETS SKIRTBOARD

CUT THE ANGLE before the skirtboard goes in. The baseboard meets the skirtboard at an angle cut into the skirtboard before it's installed. A small piece of baseboard with the same angle tests the fit.

A butt joint simplifies the fit. A glued butt joint makes fitting the base to the skirtboard a whole lot easier.

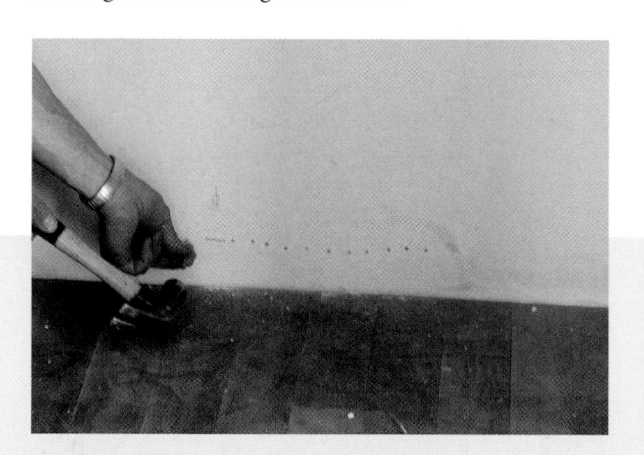

HOW TO FIND THOSE STUDS

WHEN ALL ELSE FAILS, make lots of holes. Sometimes the only way to find a stud is to make a series of nail holes in the wall (see the above left photo). When the stud is located, find the edges and mark the centerline of the stud.

Electrical boxes in new construction are also a good indication of stud location, and a rap with a knuckle shows on which side of the box the stud is running (see the above right photo). A tape measure is used to find the rest of the wall studs.

Baseboard Done Right

BY GARY M. KATZ

Carpenters new to finish work often cut their teeth on baseboard, and for good reason. Baseboard has many of the basic joints that form the foundation for trim carpentry. Over the years, I've shown a lot of carpenters better ways to run baseboard. Whether you're a veteran or new to the task, I'll share some tips that will improve both the speed and the quality of your trim work.

Measure once, measure precisely

The first key to installing any trim, especially baseboard, is recording accurate measurements on a cutlist (see the sidebar on p. 51). After years of practice, I've learned how to read a measurement when the tape is bent into a corner. But there are several other ways to measure precisely (see the photos on pp. 50–51). One method is with a measuring block. For the block, cut a piece of baseboard to an exact length that's easy to remember and add (4 in. is the length I normally use). Stick the block at one end of a run, measure to it, and add the length of the block.

For measuring to the eased edge of casing, simply lay the block flat and take a precise measurement to the crisp edge of the block. The same strategy works for an outside corner, but make sure to check the angle with a protractor to ensure that the cuts will leave you with a tight joint.

(Continued on p. 53)

MEASURE EACH LENGTH PRECISELY

Accurate measurements translate to tightly fitting trim. If you're not comfortable reading a tape measure bent into a corner, the techniques shown here yield exact dimensions.

A BLOCK ELIMINATES GUESSWORK. A measuring block makes it easy to get exact **inside-corner** measurements. Measure to the mark left by the block, then add 4 in.

EASED EDGES OF CASING CAN BE TOUGH TO MEASURE TO, so lay a block flat and measure to the edge.

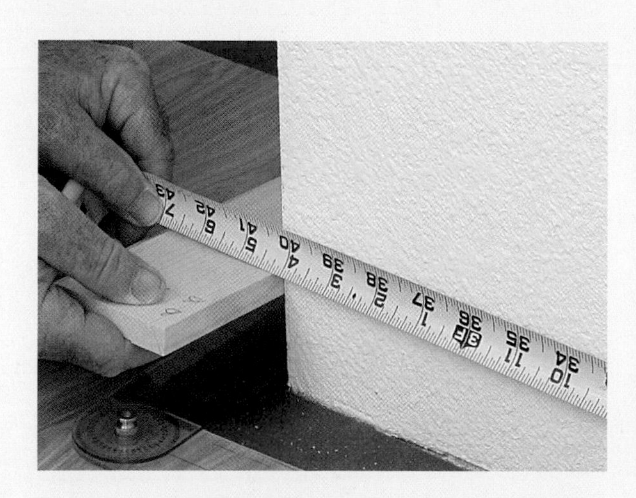

ACCURATE ANGLES AND EDGES. The rounded edge of drywall corner bead can make measuring an **outside corner** difficult, so a block works well.

OUTSIDE CORNERS AREN'T ALWAYS 90°, so be sure to check the angle of the corner with a protractor.

HIGH-TECH MEASURING OPTION. Whenever I measure long lengths, I grab my laser measuring tool instead of a tape. Despite being bulky and heavy, this highly accurate tool is perfect for measuring into tight corners, and it works great for other trim details as well.

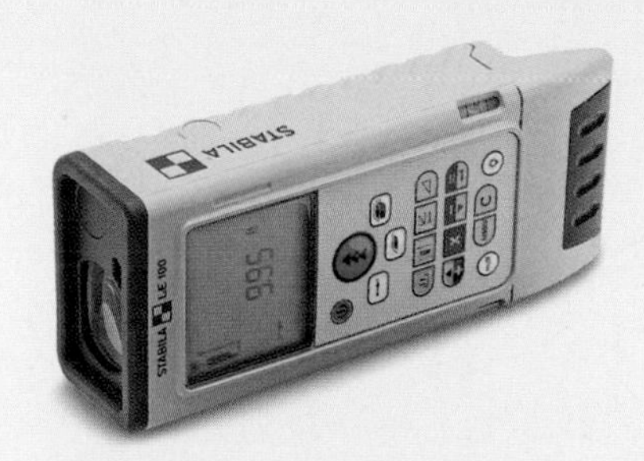

LASER® DISTANCE MEASURING TOOL
Stabila LE 100
www.stabila.com

THE CUTLIST SHORTHAND SHORTCUT

Whether it's printed on paper or on a scrap piece of wood, an accurate cutlist is the work order you take to your chopsaw. Start at a door and work around the room counterclockwise, recording the measurement for each piece and the type of joint for each end. If possible, keep most of the copes on your right side for easier right-handed cutting.

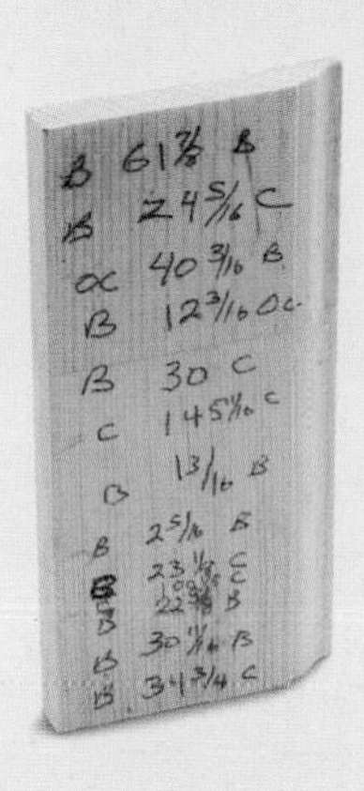

B = BUTT JOINT
C = COPE
OC = OUTSIDE CORNER

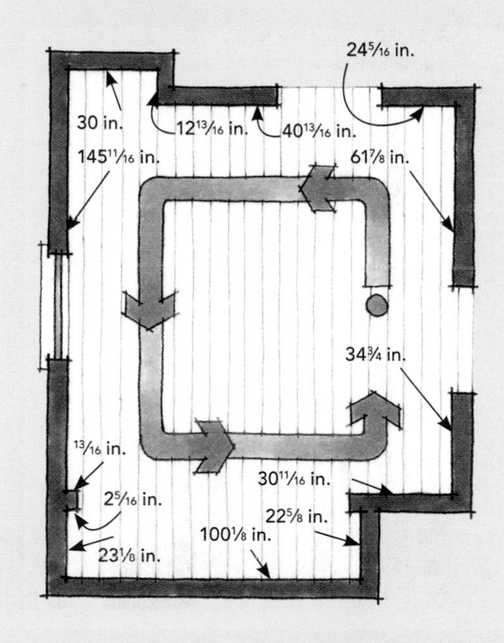

MEASURE AND CUT THE STOCK

The next step in precise trim installation is transferring the numbers from the cutlist to the stock. An auxiliary fence on the chopsaw comes in handy at this stage.

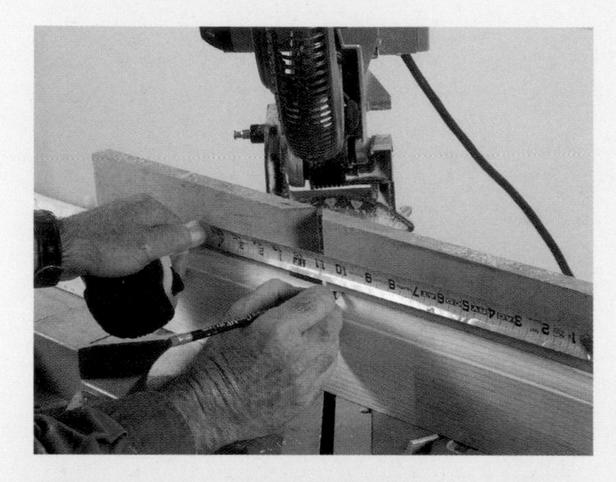

MEASURE FOR OUTSIDE CORNERS. For outside corners, the short point of the miter is always against the fence (see the near left photo). After making the miter cut, align the short point with the end of the saw fence, then hook your tape on the fence to take the measurement (see the far left photo). Spring-clamp longer boards to the saw fence before measuring.

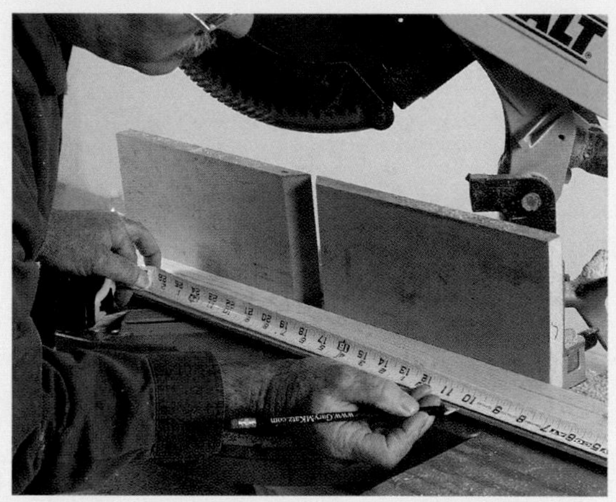

MEASURE FOR INSIDE CORNERS. For inside corners, the long point is always against the fence (see the near left photo). When pulling a measurement from a square cut or an inside corner, lay the piece flat to keep the tape from slipping off the end (see the far left photo).

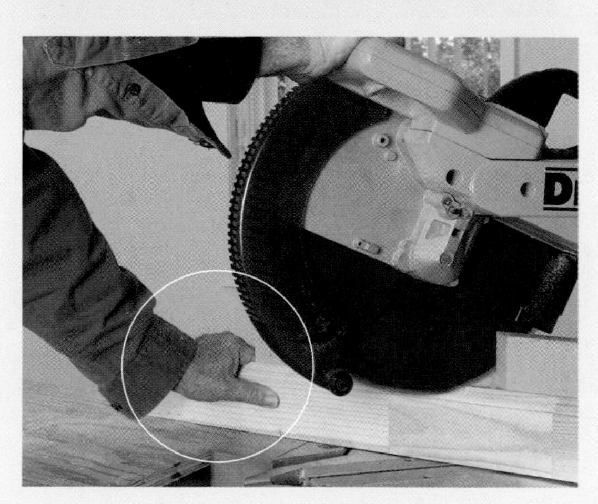

MICROADJUST YOUR CUT. Guide the cuts accurately by placing your fingers against the fence and your thumb on the face of the board (see the drawing below). Creep up on the exact cut by making incremental adjustments with your thumb (see the photo at left).

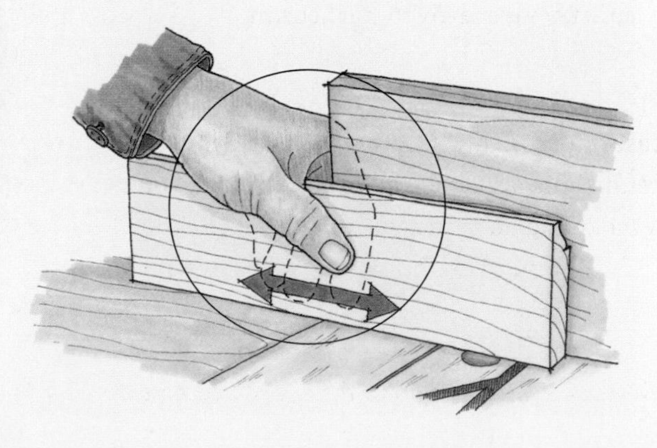

My rules for measuring change slightly depending on the length of the piece. For pieces longer than 6 ft., add about ⅛ in. so that you can spring the piece into place for a tight fit. For pieces shorter than 6 in., subtract a little so that the piece slips in easily between the casing and the corner. For everything else, measure precisely to the nearest $1/32$ in.

Cutting corners made easy

Once you've recorded all your measurements on a cutlist, it's time to head for the miter saw, which, by the way, should never be on the floor. Working on a stand with support for the work on each side of the saw is more efficient and more accurate. Many commercial stands have built-in wings, or you can go for a homemade version. Whatever the strategy, the wings should support the work so that it stays flat on the saw.

I also attach a straight, flat auxiliary fence to the saw. This added fence supports the trim as it's being cut and serves as a measuring aid, too. Make the auxiliary fence the maximum height your saw will cut, or at least as tall as the trim you're cutting and slightly longer than the fence on the saw.

You made the cutlist so that you don't have to visualize the joints as you cut them. Now it's important to teach yourself which way to swing the head of the saw to make a cut for a corner. Instead of visualizing the joint or the location of the baseboard, you can memorize two simple rules: For inside corners, the long point is always against the wall and the fence (see the photos on the facing page); and for outside corners, the short point is always against the wall and the fence. Armed with these two rules and the cutlist, you're ready to start cutting.

The fence helps with measuring

Measuring from the butt end of a board is easy. After making a square cut, lay the board flat on the saw table, and hook the tape on the end. Laying the board flat minimizes the chance that the end of the tape will slip off.

Measuring for an inside-corner cut is also simple. Make the 45° mitered cut for the coped end first. Again lay the board flat and face down, and this time, hook your tape measure on the long point of the cut.

For measuring outside-corner cuts, I put the auxiliary fence to use. Instead of "burning an inch" (measuring from the 1-in. mark), align the short point of the cut with the end of the saw fence, and pull the measurement from the edge of the fence. I spring-clamp longer pieces to the fence before measuring, or I flush the short point with the end of the extension wing, then hook my tape on the extension wing to pull the measurement.

I always use a sharp $2^5/_{10}$ pencil to mark measurements. The harder lead keeps the pencil sharper longer, and crisp, clean lines are essential for accurate cutting.

Controlled cuts

You're finally ready to cut. Just align the blade on the line and cut away, right? Wrong! It's imperative that the blade cut exactly at the line, and absolute control of the material and saw is the only way to ensure an accurate cut.

To control the material, keep your free hand locked behind the auxiliary saw fence, with your thumb over the material. Begin the cut with the measurement mark ⅛ in. or so away from the blade. Now raise the blade, and using your thumb as a micro-adjuster, slide the material over slightly and let the blade enter the material again. Repeat the process until the blade is exactly at the line, and only then cut through the piece.

A jigsaw makes the best copes

Here's where I generate all those nasty letters to the editor: Coped corners are faster and more accurate than miters (45° angle cuts on both pieces). For a crisp joint line, miters have to line up exactly in a corner. If the coped piece is off plumb slightly or not quite square to its neighbor, it's harder to detect. The extra couple of minutes needed to cope more than makes up for the fussing that most miters require for a good-looking fit.

I still use a coping saw if I'm installing only a piece or two. The key thing you have to remember when using a coping saw is to keep the sawblade moving. Don't try to cut quickly, and don't push too hard on the blade. Instead, move the saw at a comfortable machinelike rate that you can maintain throughout the entire cut.

I've recently switched to using a jigsaw for copes, but it's equipped with a Collins Coping Foot (www.collinstool.com). The coping foot replaces the base, or table, of a jigsaw, and its shape allows you to follow the profile of a molding from the back side of the material while keeping the saw base in contact with the material.

Be sure to use the right blade with the jigsaw. Collins recommends a Bosch® 244D blade (www.boschtools.com). With only 6 teeth per in. and deep gullets, the blade cuts aggressively and clears the kerf of waste quickly. The teeth also have a wide set, so the blade cuts a wide kerf, which allows you to scroll-cut almost any profile. The wide set of the teeth means you also can remove material with the side of the blade to tune the cope.

The step-by-step process of coping with a jigsaw is much better illustrated visually (see the photos on the facing page). In simplest terms, secure the board, and begin the cope with a series of relief cuts at critical points on the profile. For the colonial baseboard pictured here, I made two cuts. I first cut the sliver that forms the miter at the top of the intersecting pieces—the one that always breaks off before you can get the board nailed in. (I don't make those little slivers wafer thin anymore. Instead, I leave them at least ⅛ in. thick, and then I mortise the intersecting piece of baseboard to accept the overlapping sliver, as you'll see in the next section).

The second cut is to the point where the plinth, or the flat section of the baseboard, meets the curved profile. After making the relief cuts, it's just a matter of following the profile with the saw held at a slight angle to create a back bevel.

Installation is the easy part

After you've run through your list and cut and coped every piece, spread the pieces around the room close to where they're going (see the photos on pp. 56–57). I generally install baseboard in the same order that I measured and cut it, except that I insert any small pieces (up to about 2 in. wide) first, such as between casing and a corner. Don't try to nail these small pieces, or they'll split. Instead, let the intersecting board hold them in place.

For this type of baseboard, which is only 3½ in. tall, don't bother trying to locate the studs. The wall plate is usually high enough for me to hit with 2½-in. finish nails. For those areas where I need to nail higher, I simply find a stud with a magnet.

In a coped corner, always install the butt end piece first. Then slip in the coped piece. Pressing the coped piece tightly to the wall, plunge the blade of a utility knife into the top of the butt end using the edge of the miter cut as a guide. Remove the coped piece, and cut a notch deep enough to accept the overlapping sliver on the coped end.

Now the coped piece can go in permanently. If your copes are accurate, the fit will be tight. Tap the butt end of the coped piece with a block and hammer until the two pieces marry tightly in the corner and the butt end of the coped piece slips into position.

For outside corners, always test-fit the pieces before nailing them in. When you're satisfied with the joint, apply glue to the mating surfaces of both pieces. Wipe off the excess glue, taking care not to get glue on the baseboard's finished face. The glue joins the two pieces and also seals the ends of the boards. For stain-grade installations, sealing keeps the ends from absorbing more stain and becoming darker than the rest of the board. I preassemble outside-corner joints using a 23-ga. pin nailer that leaves almost invisible nail holes. Then I nail the whole assembly into place.

COPING WITH A JIGSAW

AN ODD-LOOKING JIGSAW BASE called the Collins Coping Foot [1] allows you to cope quickly and accurately. Be sure to use an aggressive blade (6 tpi) with a wide set. Pull the saw toward you to make a relief cut at the top of the miter, [2] leaving the sliver big enough so that it doesn't break off easily. The next relief cut [3] is where the profile begins. Keeping the saw at an angle, follow the top profile [4]. Switching to a push stroke, cut the flat part of the profile, [5] or the plinth.

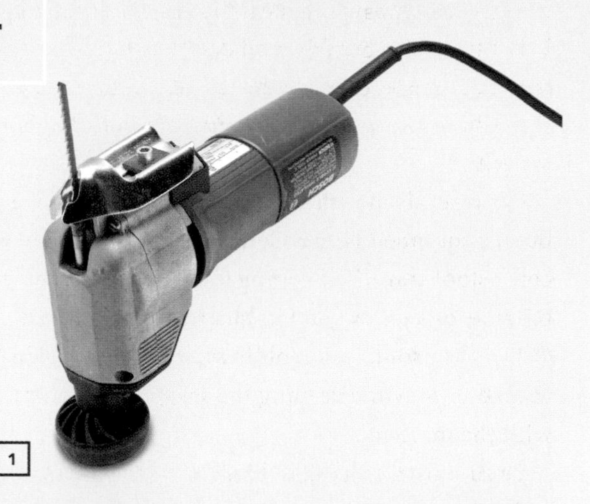

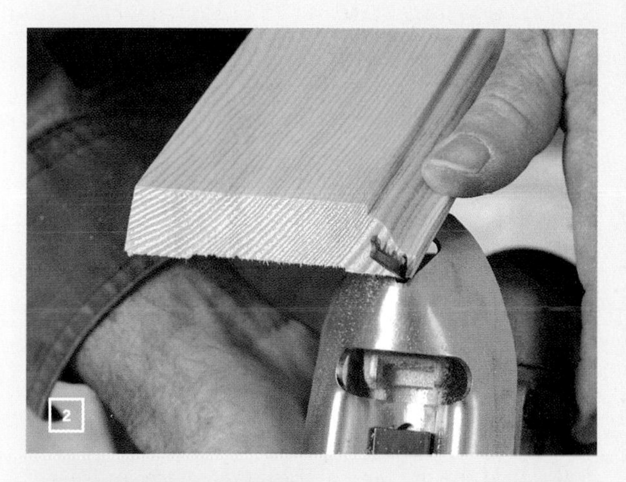

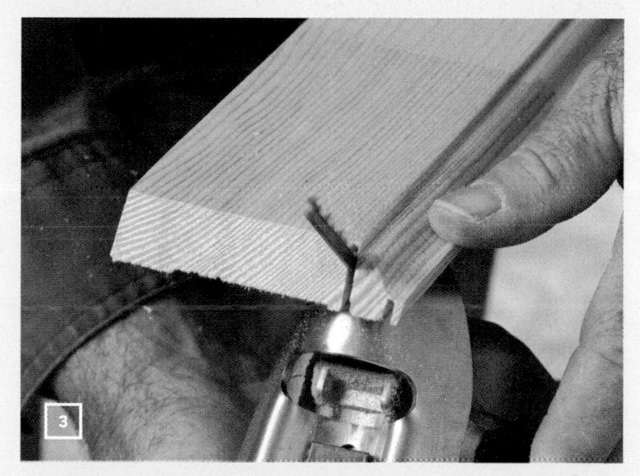

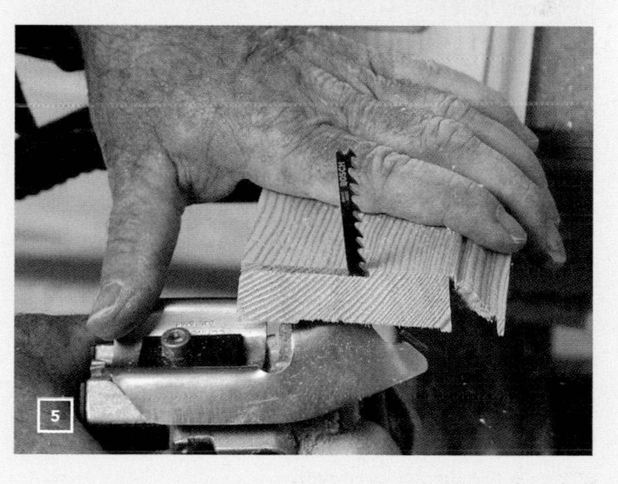

INSTALL THE PIECES IN ORDER

Spread the cut pieces around the room close to where they will be going. Then work around the room in roughly the same order that you took the measurements. For inside corners, always install the square-cut or butt-end corner pieces first.

DISTRIBUTE THE CUT PIECES. Place the lengths of baseboard around the room near where they are going to be installed.

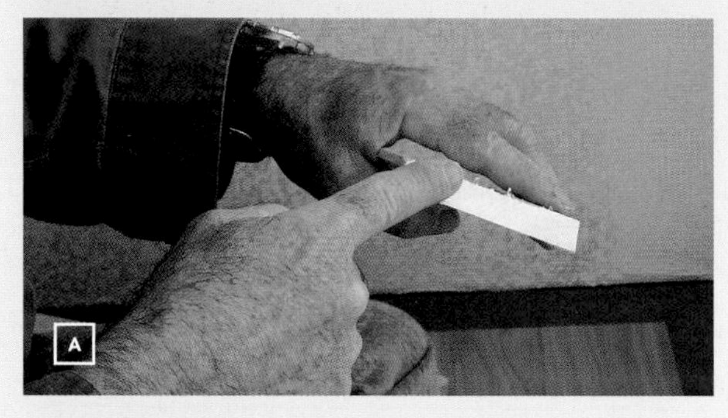

GLUE UP AN OUTSIDE CORNER. For outside corners, spread glue evenly along both mating faces. Assemble the joint with 23-ga. pin nails.

FITTING SMALL CORNER PIECES. Small pieces go in first and are held in place by adjacent pieces.

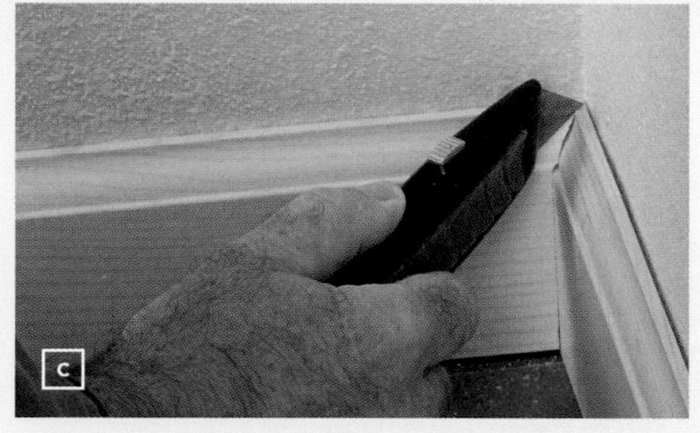

FITTING AN INSIDE CORNER. To fit a coped corner, line up a utility knife along the edge of the miter cut (see the photo above), and notch the butt end slightly for the top of the miter (see the photo at right). For tight-fitting pieces, tap the coped piece with a hammer and a block until it slips into place.

Keep baseboard straight when the walls aren't

As you can see from my cutlist, I sometimes like to cope both ends of longer pieces. With the little bit of extra length I give them, they spring into place with tight pressure-fit joints at both ends.

Most carpenters I know just nail the board tight to the wall. If there are any bows or bellies in the wall, though, making a piece of trim conform to them only accents the flaws (see the photos on p. 58). It's a good idea to check long pieces with a string before nailing them in. Shim the baseboard so that it looks straight to the eye, then caulk any gaps between the top and the wall. The straight baseboard will make a crooked wall look straighter.

Preassemble all short sections, such as around a column, around an end wall, or for a drywall return at a door opening. Again, if the two sides aren't square to each other, installing an out-of-square joint just makes the return look worse. Instead, make the joint square, shim the short piece, and caulk the top edge.

Finally, in areas where drywall is broken or missing, insert a drywall screw to keep the baseboard straight and plumb. Simply adjust the screw until the head is flush with the surrounding drywall. The screw head backs up the base to keep the joint looking good while the baseboard hides the drywall defect.

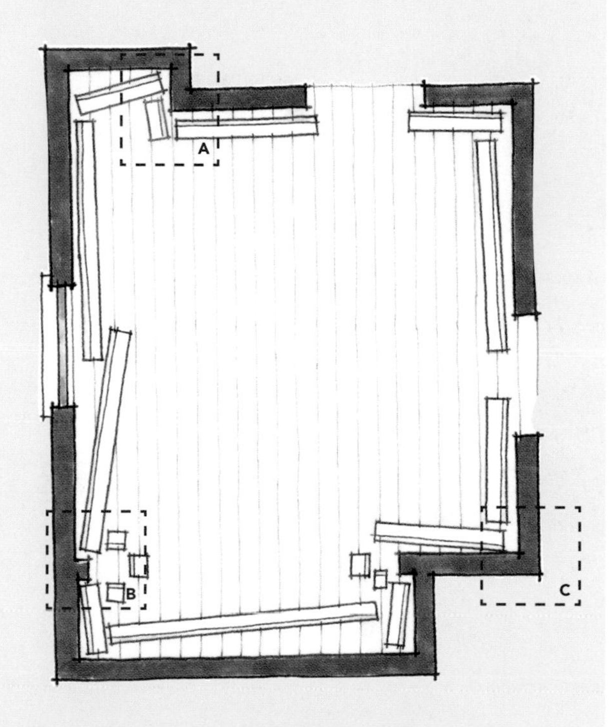

TIPS FOR THE PROBLEM SPOTS

Keeping baseboard straight will improve the look of a crooked wall or an out-of-square corner. When drywall is damaged or missing, keeping the board plumb can be tough.

SHIM LONG BOARDS. Longer boards spring into place for tight joints (see the above left photo). Check them with a string, and shim them straight, if necessary (see the above right photo).

SHIM CORNER RETURNS. Shim corner returns square (see the above left photo), and caulk the gap along the top (see the above right photo).

MISSING DRYWALL IN CORNERS. Where the drywall is missing, insert a drywall screw to keep the base plumb in the corner (see the above left photo). Adjust the screw until it's even with the wall (see the above right photo).

Crown-Molding Fundamentals

BY CLAYTON DEKORNE

The first time I installed crown molding, I couldn't believe how easy it was. The builder I was working with knew exactly where every stud, truss chord, and piece of blocking in the condo complex was located. The walls were flat, the corners square, and the ceiling level; there were no outside corners and no extra long runs. I wouldn't find out that I'd been lucky rather than smart until a year later when I tried to put up crown in an old Victorian town house, where nothing was plumb, level, or square, and it seemed as if no framing existed beneath the plaster.

Angled between the ceiling and the wall, crown molding inhabits a three-dimensional space, making it one of the most demanding types of trim to install. When the wall turns a corner, crown makes two turns: One is along the ceiling plane, and one is along the walls, requiring compound cuts to join the pieces. Any discrepancy in either plane introduces a twist or bow that can alter the angles. Crown also is affected by shifting house movements acting on both planes, so it needs to be nailed securely to a stable surface to ensure that the joints don't separate over time.

Every crown gets backing

Even when I'm following a first-rate framing crew, experience has taught me to check every corner with a framing square and to note any serious irregularities. (A gap between the square and the wall corner of even ⅛ in. can make it difficult to close the joint on a large piece of hardwood crown.) At the same time, I eyeball the wall and ceiling intersection, looking for any irregularities that might hinder the crown molding from making full contact. If I encounter a saggy ceiling or a wavy wall, I take the time to figure out if corrective measures are needed (see the sidebar on p. 67).

Even when everything is square and true, I still have to make arrangements for fastening the crown. Simply nailing into the studs and ceiling joists secures the crown only along the top and bottom edges. Inevitably, the edge splits, or the small finish nails don't find good purchase and give out after a few years of seasonal movement. I avoid such aggravations by installing a backer of rough framing lumber that allows me to nail the crown wherever I want.

For modest-size crown, my typical backer is a straight 2×4 ripped at a bevel to match the spring angle of the crown (see the top photos on p. 60). Rather than cut the backer to fill the space behind the crown completely, I lay it out and cut it so that the beveled face is about ¼ in. shy of the back of the crown. This gap allows enough wiggle room to fine-tune the joints if needed. The crown still firmly

(Continued on p. 62)

TWO TIPS TO MAKE THE JOB GO SMOOTHLY

A. ALWAYS PUT UP A BACKER

A FRAMING SQUARE AND A SCRAP of crown determine the backer dimensions. After ripping the material on a tablesaw, the author uses 3-in. screws to attach the backer to the framing.

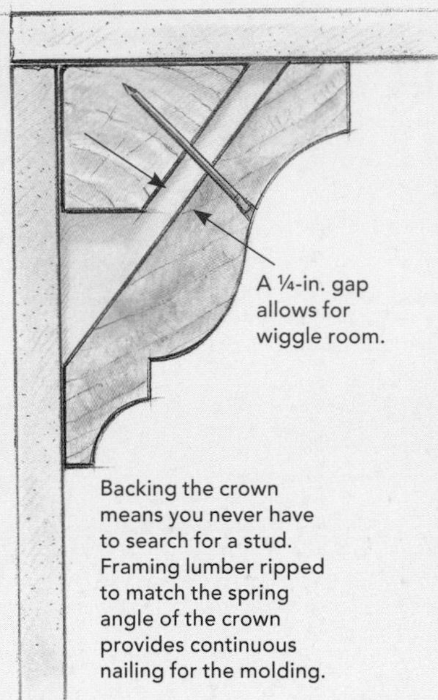

B. ALWAYS START WITH A COPED JOINT

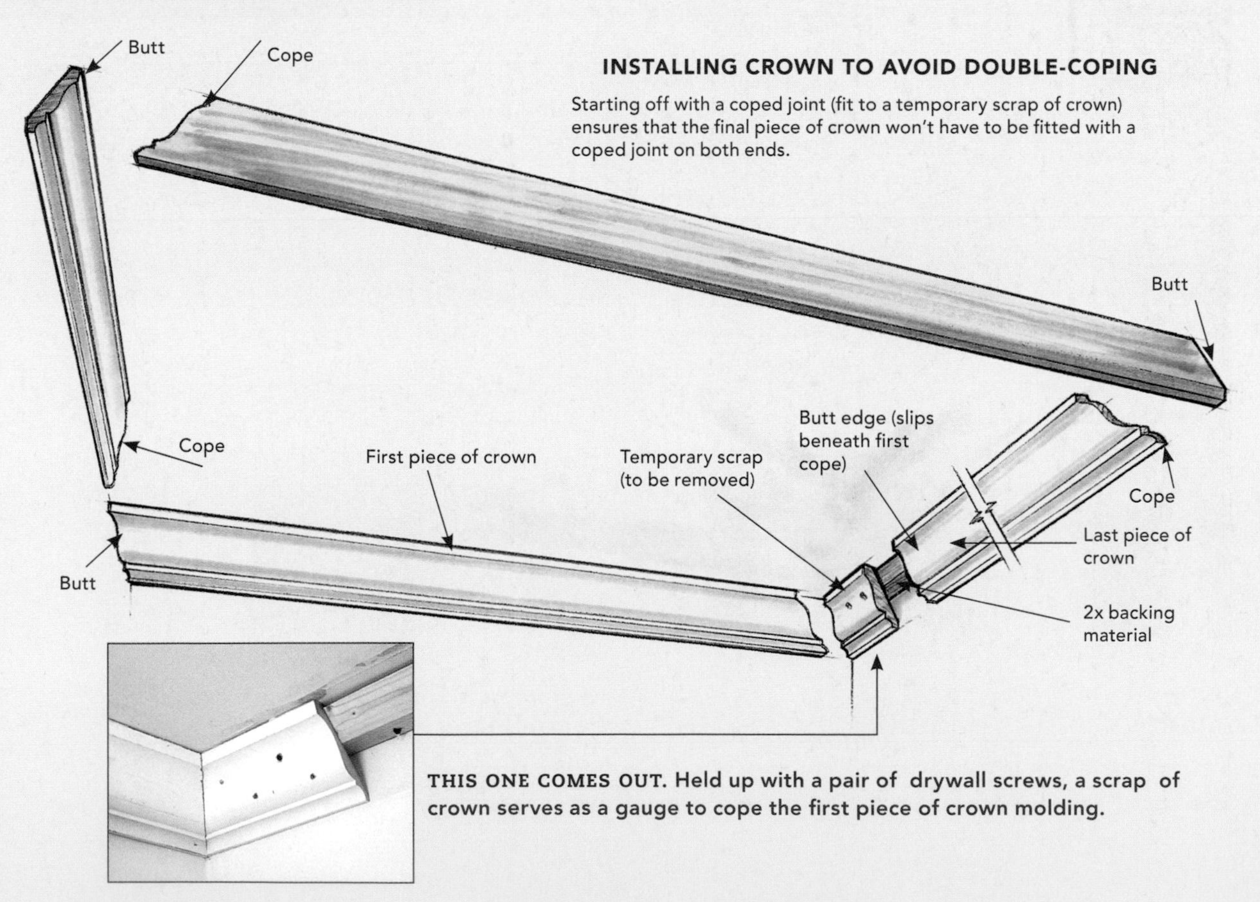

THIS ONE COMES OUT. Held up with a pair of drywall screws, a scrap of crown serves as a gauge to cope the first piece of crown molding.

B
A

contacts the ceiling and wall at the edges, but its meaty belly is pinned with a stout finish nail.

I find it's usually easier to run the backer continuously. But if I'm running low on backer material or if I want to avoid high spots in the wall or ceiling, I may run the backer in sections—whatever it takes to give me nailing every 16 in. to 24 in. On walls long enough to require two lengths of crown, I figure out in advance where the joints will be and leave about a 16-in. gap between backer sections to accommodate the method I use to preassemble scarf joints (see the sidebar on the facing page).

Plan the installation to avoid double-coping

Before I start chopping up the molding, I scope out the installation order. To avoid coping two ends of the same piece of molding and to eliminate the time and fuss of resetting my saw for each cut, I try to set up all my copes on the same end of each piece. Most right-handers find it easier to cope right, butt left. The direction doesn't matter as much as staying consistent; find what suits you and stick to it.

If the room has an outside corner, I usually start from there. Otherwise, I start with the trim on the longest wall opposite the door. I install this piece with

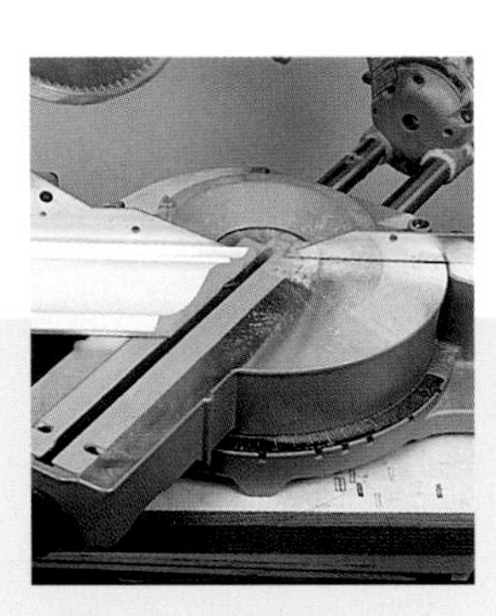

AN INSIDE-MITER CUT defines the profile of the cope.

BACK-CUT THE COPE to ensure a tight joint.

FOR BEST RESULTS, approach the cut from more than one direction.

COPING INSIDE CORNERS

IT'S POSSIBLE TO ASSEMBLE inside corners with a miter. However, even if the corner is perfectly square, most miter joints open after a few seasons of expansion and contraction. I prefer to cope crown molding. A coped joint is a marriage of opposites. On one side, the molding is cut square and butts tight to the wall. On the other side, the molding is "coped" to match the profile of its neighbor (see the above left photo).

A coped joint starts with a compound miter, the same as for an inside miter (see the middle left photo above). The contoured edge along the face of this cut defines the profile to be coped. I mark the edge of this profile with the flat of a pencil lead so that it's easier to see as I cut it with a coping saw (see the middle right photo above). I always back-cut slightly by tilting the coping saw past 90° to ensure that the two pieces of molding intersect only along the profile line. When cutting with a coping saw, I typically saw toward the curve from different directions so that I don't have to twist the saw and risk breaking the fragile edge of the profile (see the above right photo).

OUTSIDE CORNERS ARE SIMPLE UNLESS THEY AREN'T SQUARE

WHEN I ENCOUNTER AN OBTUSE or acute angle corner, I use a couple of sticks and a bevel board to measure the exact miter angle. The bevel board is nothing more than a square piece of wood with some carefully drawn angled lines (from 0° to 90°). I made mine using a simple protractor and a sharp pencil (see the photo below right). To get the exact bevel angle for an out-of-square corner, start with two short scraps of wood that overlap at the corner (see the top photo):

1. Mark the long point of the angle on the upper piece.
2. Mark the short point on the same piece.
3. Find the angle on the marked piece with a bevel square.
4. Then lay the bevel square on the bevel board and read the angle.

A more modern alternative is the Bosch DWM40L Miterfinder (see the bottom right photo). This ingenious device can measure any angle to within one-tenth of a degree; then with the push of a button, it gives you the miter and bevel angles for any crown molding.

THE MITER FORMED BY the intersection of two boards is transferred to a bevel square.

A HOMEMADE BEVEL BOARD reveals the angle of the miter.

MARKING THE LONG POINT is generally more accurate than marking the short point. To establish the long point of an outside miter, trace the top edge of a scrap of crown onto the ceiling, on both sides of the corner, then measure to the intersection of the two.

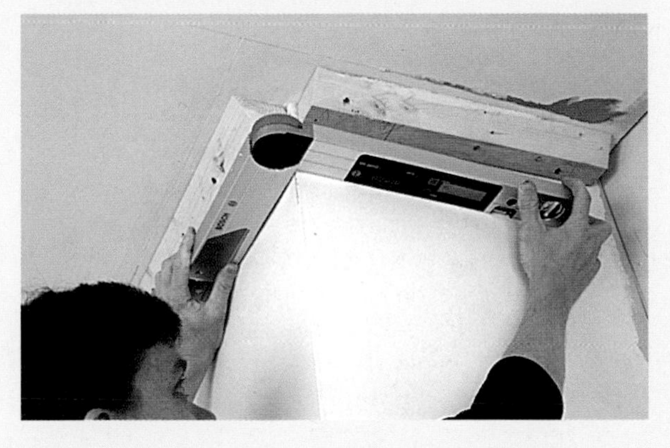

FOR ABOUT $110, a digital protractor does the thinking for you.

one coped end (see the sidebar on p. 62) and the other end butted square into the corner.

When I set up all the cuts in the same direction, the coped end of the first piece of the molding must be installed against a temporary block (see the photo on p. 61). This block is a scrap piece of crown, about a foot long, that is screwed temporarily to the backer. I nail the first piece of crown near the center and toward the butt end only. This placement gives me enough support to hold up the entire length of crown, but it leaves the coped end free so that I can slip the last full-length piece of crown in place of the starter block when I finally get there (see the drawing on p. 60).

Pinch sticks make for easy, accurate measurements

Because joint compound tends to build up in the corners, it's hard to get an accurate measurement by bending a tape measure into a corner. For long lengths of crown, I add a heavy $1/16$ in. to all my measurements; then I put a slight back-cut on the square end. That way, even if the piece of crown is slightly long, it can be sprung into place, and the point at the crown's bottom edge digs into the wall.

To measure for short lengths of crown, I always use pinch sticks, butting one end to the previously installed length and the other to the opposite wall (see the photo below). By holding the two sticks together and sliding them until one end of each hits the wall, I get an accurate representation of the distance that can be transferred easily to the molding. Once I have the distance, I mark the sticks with two lines for positive realignment, and I use a spring clamp to hold the sticks together when transferring the length to the molding.

Outside miters must fit precisely

If the corner is crisp and square, I measure the length for an outside miter by holding the stock in place and marking the bottom edge of the molding where it meets the outside corner. (Bear in mind that

PINCH STICKS YIELD PRECISE MEASUREMENTS

SLIDING TWO STICKS AGAINST OPPOSITE WALLS and marking them after each makes contact is an effective method for measuring between walls. The ends of the pinch sticks are tapered to make one positive contact point.

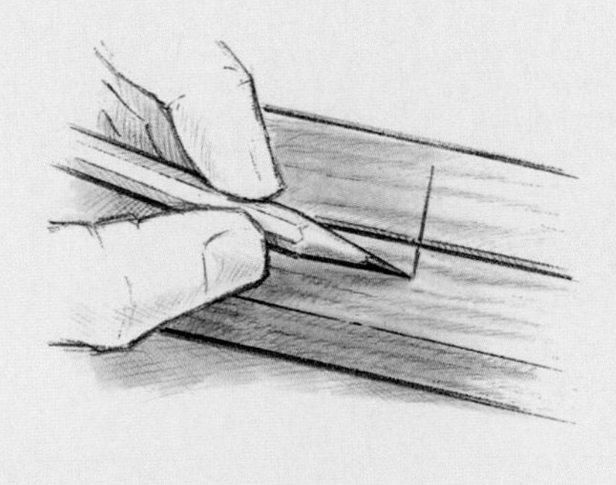

WHAT TO DO WHEN THE WALL IS LONGER THAN THE STOCK

IF A WALL IS LONGER than the crown stock available, you need to splice (or scarf) two lengths together. Rather than attempt to fashion a good fit on the wall, I create my scarf joints first.

For scarf joints, I depend on two critical devices: a biscuit and a splint. I begin by cutting a 20-degree bevel and a 20-degree miter—or 20/20 compound cut—on one end, and a corresponding 20/20 compound cut on the mating piece. I prefer a compound cut because it's easier to hide the joint; to make this cut on a standard miter saw, use the jig as shown in the bottom left photo on p. 66. With the molding and the biscuit joiner lying flat on the same surface, I make biscuit slots in each piece [1].

The slot should fall in the center thickness, not through the contoured face of the molding. The long point of the bevel on the compound cut holds the biscuit joiner away from the surface in which the slot is cut, but by setting the biscuit joiner to a maximum depth, I can fit a smaller #0 biscuit into the slot.

To fasten the two pieces more securely, I cut a strip of ¼-in.-thick lauan to act as a splint across the back and to bridge the joints. I fasten the splint with plenty of glue, using ½-in. staples and a standard staple gun to hold the strip until the glue cures [2]. After securing the splint to the scarf joint, I smooth out the joint with sandpaper [3]. Until they're securely fastened in place, I make sure to handle these scarfed pieces with extra care; they can snap in two easily.

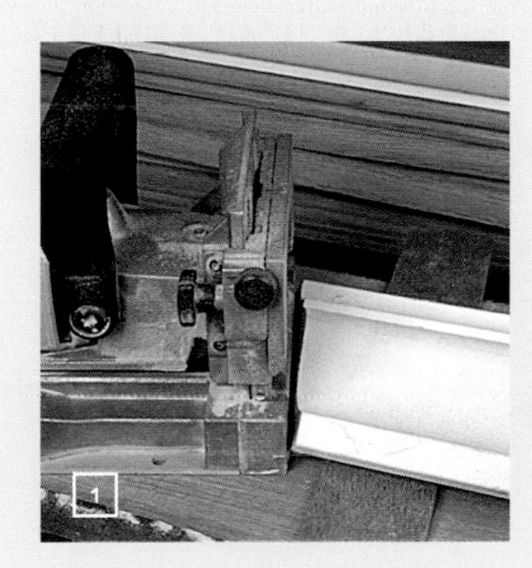

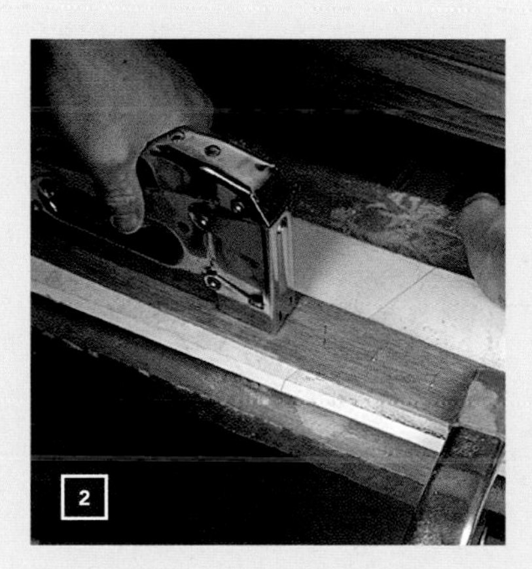

drywall corner bead pushes the corner out slightly, so you may have to back-plane the miter to get a tight fit.) Plaster corners may be rounded slightly, making it hard to measure the exact length along the crown's bottom edge. If the walls are reasonably square to each other (and only the corner is rounded), I mark the ceiling by tracing along the top edge of a piece of molding, first along one side of the corner, then along the other. Where these lines cross on the ceiling marks the long point of each miter cut (see the bottom left photo on p. 63).

If you have to assemble more than one outside corner, consider cutting them all at once. I still try to map my cuts so that all the cuts for the copes angle the same direction. Rooms with bump-outs often have a few small pieces of crown. Wherever a small piece must be coped on one end and mitered on the other, I cope the piece before I cut it to length. In this case, I might have to tack up the coped length of the outside corner temporarily so that it can be removed later to fit the last piece of crown that laps behind it.

Dual compound-miter saw makes cutting crown less confusing

If there's any secret to cutting crown, it's simply this: Keep track of which edge of the molding is the bottom and which edge is the top. Doing so eliminates a lot of head-scratching when you're standing at the saw trying to remember how the molding will sit when actually nailed into place. I try to group similar cuts together. That is, I first cut all the scarf joints, then cut all the outside-miter joints, then move on to the inside-corner cuts. This strategy saves time with any trim. But with crown molding, it's especially helpful because it lets me keep track of the three-dimensional orientation of the molding.

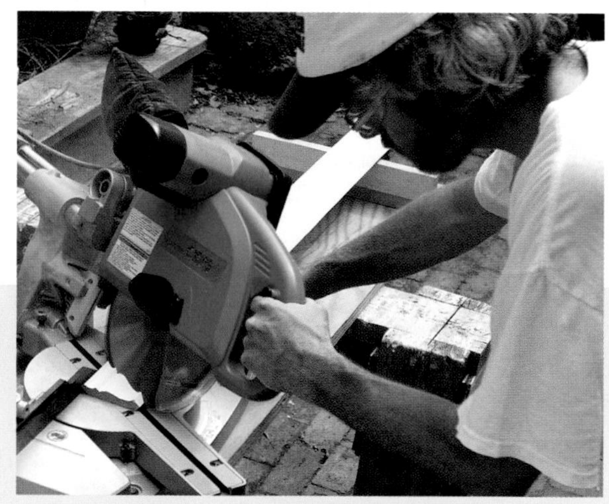

BUILD A JIG OR BUY A NEW SAW?

IF YOU HAVE TO CUT CROWN on a standard miter box, an auxiliary plywood fence with a stop applied to the front edge ensures that the molding is held at the correct angle in relation to the blade (see the above left photo). Because the saw head on a dual compound saw tilts in two directions, you never have to flip the molding around to set up the correct cutting angle, which is a big advantage when working in tight or crowded quarters (see the above right photo).

Keep track of which edge is the bottom and which edge is the top.

The methods you use to cut inside and outside corners depend on the type of saw you're using. With a conventional miter saw—and narrow crown—I use a simple jig that holds the crown in the saw at the precise angle in which it will be installed.

Unless you've got a huge miter saw, such as the 15-in. Hitachi® C15FB, you need to use a sliding compound-miter saw if the crown is wider than 5½ in. A sliding compound-miter saw allows you to cut the crown flat on the saw table. With most sliding miter saws, the motor head pivots in only one direction; this means that you have to line up either the top or the bottom edge of the molding on the fence to get the cut in the right direction, so you wind up flipping material a lot.

Recently, I began using one of the new dual compound-miter saws. I admit I was skeptical at first that such a big lug of a saw would offer much of an advantage—unless I was dicing up huge pieces of lumber. But a dual compound-miter saw tilts in two directions, so you never have to flip the molding around. This is a big plus in a tight workspace.

With the dual compound-miter saw, I always keep one edge against the fence and move the saw head to the left or right depending on whether I am cutting an inside or outside miter, left or right.

With any compound-miter saw, you still need to find the correct miter and bevel settings. These settings change, depending on the spring angle of the crown molding you're installing, and there's no obvious correspondence between all the angles. I always keep handy a chart that lists the angle settings for the most common spring angles. For moldings that aren't included on the chart, I use the Bosch Miterfinder (www.boschtools.com) to calculate the angle settings, or I simply make a reasoned guess and cut a couple of test pieces until I get the angles right. Then I write those numbers down so that I won't forget them.

SAGGY CEILINGS AND WAVY WALLS DEMAND SPECIAL ATTENTION

In old houses, where walls and ceilings aren't smooth, I begin by mapping the planes with a level and string to find the low points on the ceiling. Long lengths of crown are surprisingly flexible and may move with gentle bumps and shallow dips. Most of the time, however, some remedy is required.

One way to solve this problem is to scribe the top edge of the molding to the ceiling. When scribing, I don't want to bite too deeply into the top edge, or the scribe will be easily visible. Generally, I don't remove more than a third of the top reveal.

Another fix is to skim-coat the wall or ceiling surface with plaster or joint compound to conceal gaps after the crown has been installed. In this case, I install the crown in line with the high spots. To float the mud, I make a screed from a short piece of mangled crown molding that has a notch in it that keys with the edge of the installed crown molding (see the photo below). This screed allows me to apply the mud to a depth that precisely covers the gap. I usually make the screed about 18 in. long, something that gives me a long, thin taper of mud that seamlessly feathers into the existing surface.

HOMEMADE SCREED LEVELS the peaks and valleys. Rather than filling gaps in wavy walls with great gobs of caulk, a notched screed, made from a scrap of crown molding, enables the drywall finisher to feather those gaps back into the wall, where they belong.

Inside Crown Corners

BY TUCKER WINDOVER

I can still remember my first day on a finish-carpentry crew. I spent most of my time keeping one eye on the lead carpenter, trying to pick up as many new tricks as I could. I watched him cope and fit a couple of pieces of crown molding. His movements were fluid. His process looked easy. But I didn't quite understand why we had to complicate a joint that could be made with two simple miters. I asked if it would be OK to miter the corners instead of coping them, and the lead said, "Go for it—if you can make it look right and convince me that it will stay that way." So I attempted to do it the easy way, or what I thought would be the easy way. In truth, I didn't make it through a single room before turning to a coping saw.

I've learned a lot since then, and in my opinion, a coped joint is preferable to a miter in several ways. Due to wood movement, miters tend to open at the short point or the long point. A coped corner stays tight as each piece of trim expands and contracts in sync. A coped joint also fits well when installed in a room with minor imperfections, such as one with corners that are out of square or ceilings and walls that are less than straight, level, and plumb. The smallest bump of joint compound can throw a mitered corner out of proper alignment.

A coped joint is also more adjustable. If there is a gap in the joint, you often need only to pry the other end of the trim to squeeze the joint tight. Finally, the coped joint is much more forgiving. This makes it a faster joint to make because there is less fussing over the fit. If a piece of trim is cut slightly short, say by ⅛ in., the gap is covered by the subsequent piece of coped trim. When you're mitering a corner, any error in length will be revealed.

Still, the coped joint has limitations. Certain crown profiles can't be coped. In these instances, I create a hybrid joint by coping what's copable and mitering what's not. When working with medium-density fiberboard (MDF) trim, I miter the inside corners instead of coping them. MDF trim is more stable than wood, but when coped, its edges become too fragile. I miter crown when installing it on kitchen cabinets that I know to be straight, level, and square. Also, I find that when preassembling crown for a column or a fireplace mantel, mitering works best.

Some people believe that coping is difficult or time-consuming. I can assure you that with a little practice, coping becomes as easy and as fast as tying your shoes.

COPING CROWN MOLDING is more than a blind following of tradition. Understand the benefits, and learn how to cut copes right.

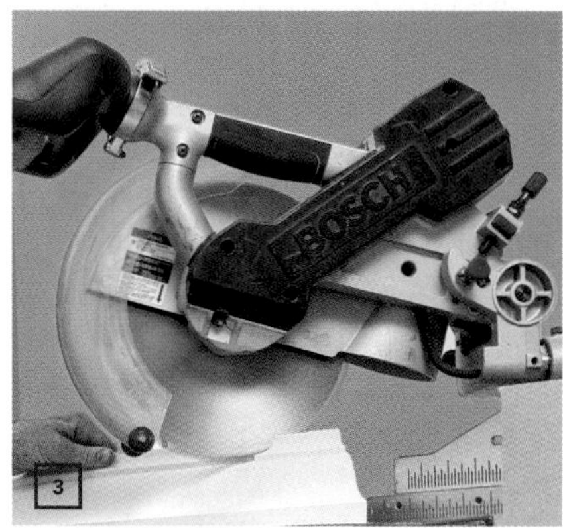

Miter the stock

To start, make a marking jig (see photo 1). Label the top of a rectangular scrap of ½-in. plywood as ceiling and its right side as wall. Measure the molding's ceiling projection and its drop down the wall by nesting it in a framing square. Transfer these measurements to the piece of plywood, and cut between the points, removing the jig's corner. This jig will be used to ensure proper crown positioning throughout the cope and the installation (see photo 2).

Create a stop that replicates the ceiling position. The distance between the stop and the miter saw's fence should match the ceiling projection. Check the stop location with the marking jig. This setup guarantees that the molding is cut in the same position that it will be installed.

Cut a 45° miter, and measure for length (see photo 3). Place the top edge of the molding down against the stop, and put the bottom edge of the crown against the saw's fence. If coping to the left, the long point of the miter will be at the bottom edge of the crown (see the photo on p. 69). Make the miter. Then, using a measurement taken between the corners of the wall, measure from the long point of the miter to mark the trim's length.

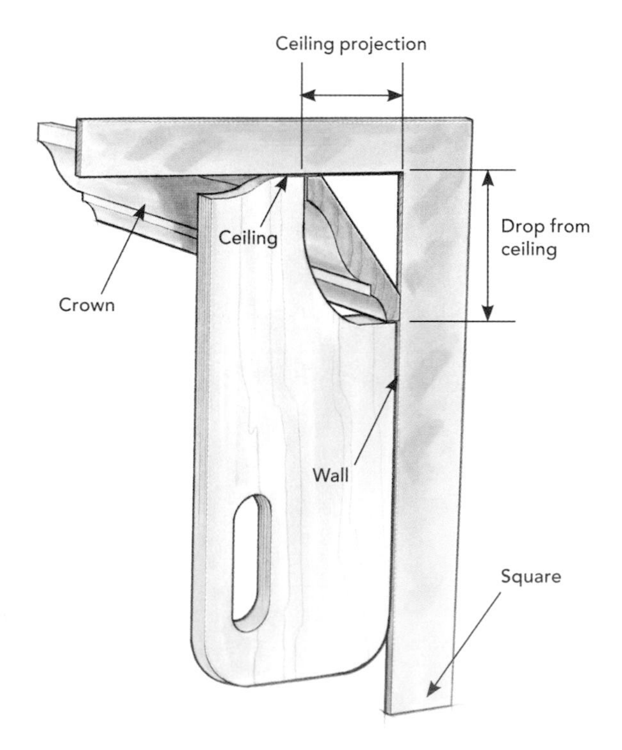

Cope the cut

An auxiliary fence helps to secure the trim for the cope. With the ceiling-side edge of the trim flat against the fence and the wall-side edge flat against the stand, snug up a stop to keep it in position. Use the marking jig to be sure that the trim is seated correctly (see photo 4).

A jigsaw fitted with a coping foot (www.collins-tool.com) and a Bosch T244D blade increases the speed of the cut. When used correctly, this setup provides a precise, controlled cope. If the jigsaw doesn't suit you, there is nothing wrong with using a traditional coping saw. With either tool, make several relief cuts before following the profile of the molding. Hold the saw somewhat less than vertical to give the cope a slight back cut (see photo 5).

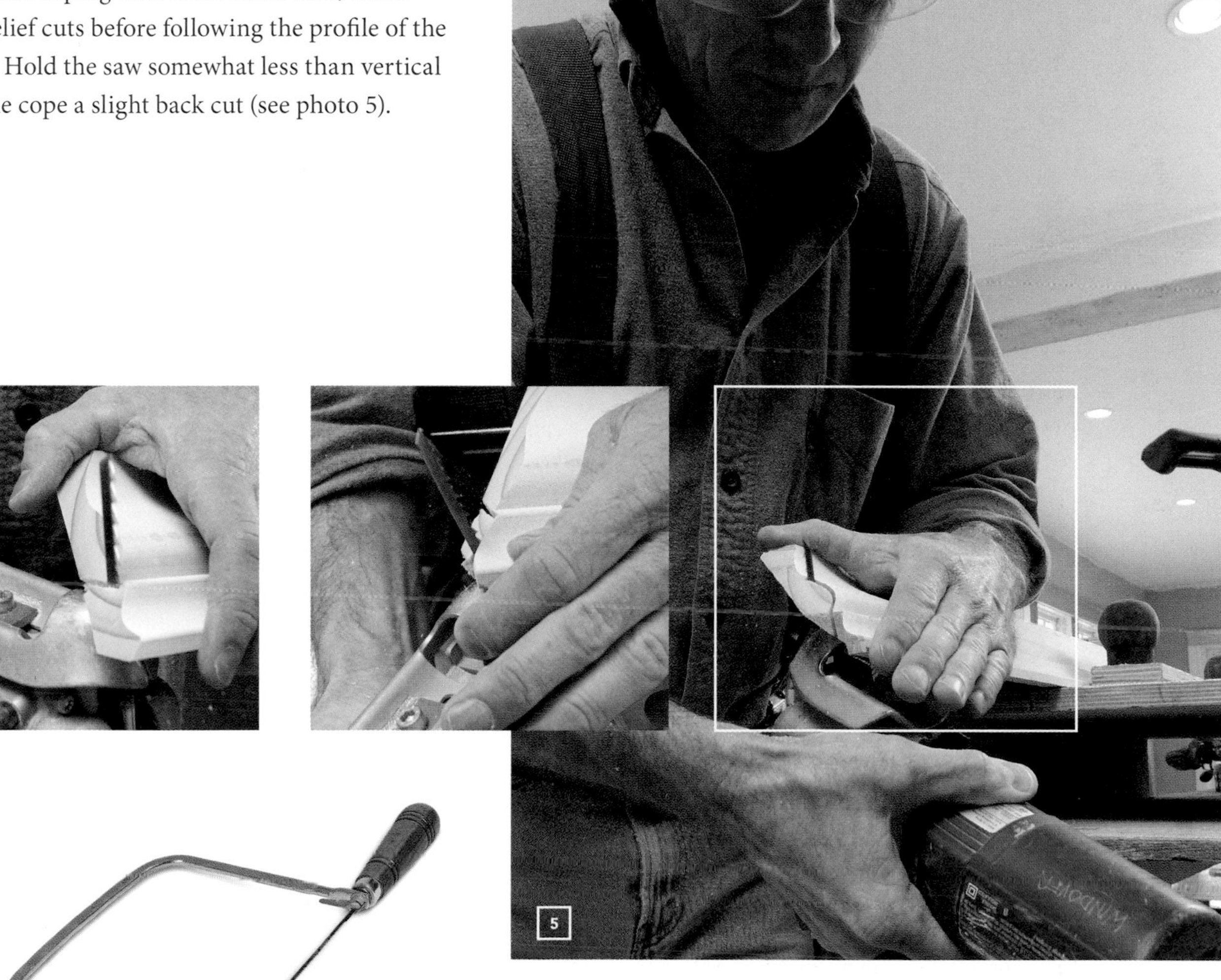

Test the fit

Use a scrap of the molding to test the accuracy of the cope. Whether the corner is less than or greater than 90°, the coped joint stays tight (see photo 6). If the joint isn't as tight as you'd like, fine-tune the cope with a rasp or a sanding sponge (see photo 6A).

Install the crown

Cut the trim to length with a 90° cut. This butt end is covered by the adjacent coped piece of trim. Cut pieces of trim more than 8 ft. ⅛ in. long. The compression helps to keep the corners tight over time. When installing the trim, don't nail within 2 ft. of each end until the joint fits the way you want it (see photo 7).

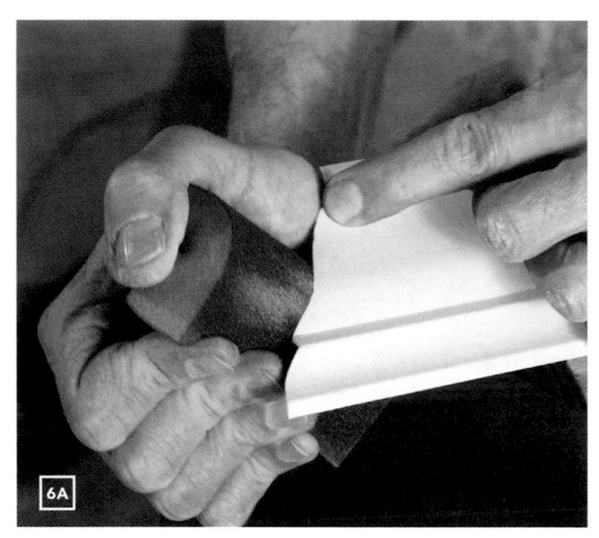

GOING AGAINST THE GRAIN: I MITER

As a student of two old-school carpenters, I was taught to cope the inside corners of crown molding. I installed my crown that way for 15 years because it was the only way the old guys would allow it to be done. However, 20 years ago, fellow finish carpenter Craig Lawrence suggested to me that it was a lot easier and faster to miter inside corners. I was skeptical, but tried that approach and found that he was correct. Since then, I have mitered my crown-molding corners with nary a complaint. With the miter approach, you simply set up some stops, pull the trigger, and slice through the stock for a perfect cut every time.

I even back-cut each piece a degree or so to accommodate out-of-square corners. If this is not done, the joint will open in front. I leave the ends of the first piece of crown molding loose until I have the second one up and in position. This allows the corner to be tweaked until it fits. I then glue the pieces together, which I have found helps to form a tight-fitting joint. Finally, I wipe the joint with caulk to blend the pieces together, using colored caulks for stained wood.

My advice? Develop a system that works for you, but give miters a try.

—Chris Whalen

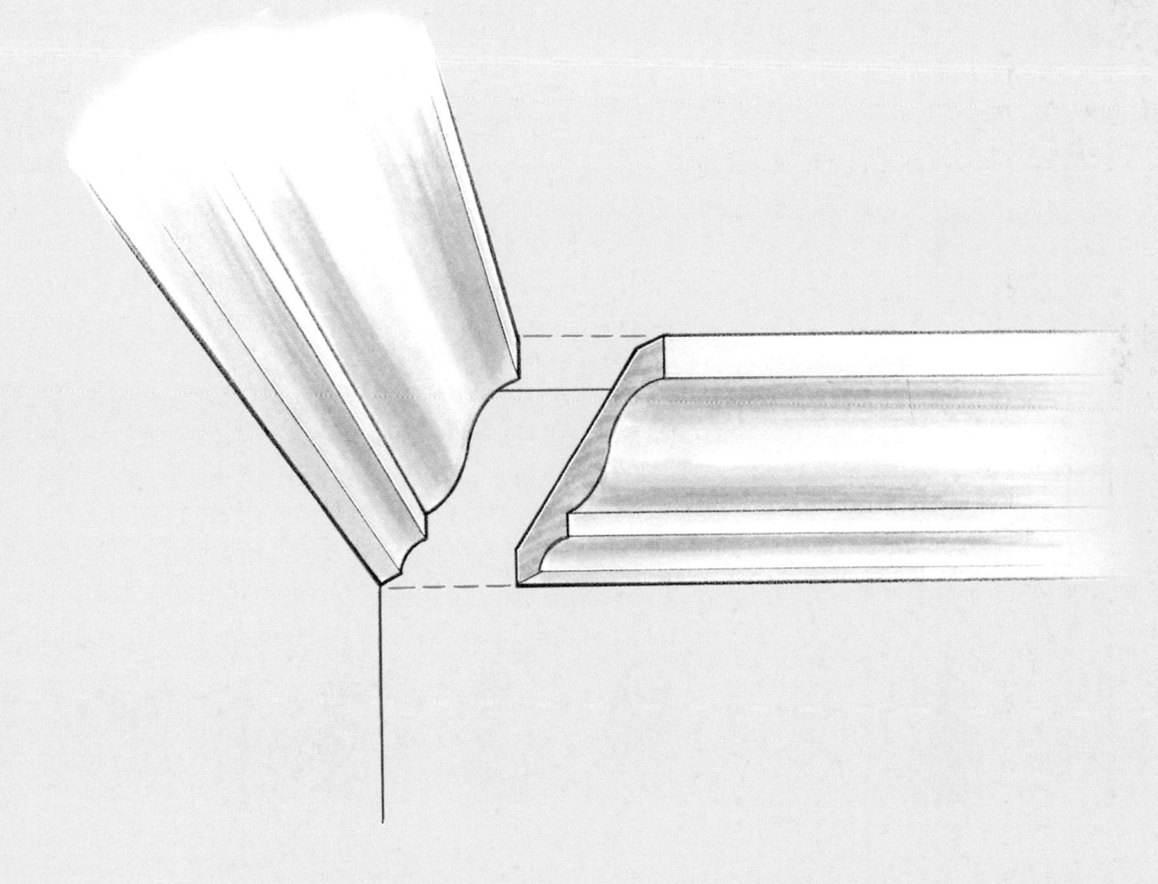

Coping Moldings

BY TOM O'BRIEN

When two pieces of trim meet at an inside corner, you could miter the joint, but most professional carpenters prefer to cope. An airtight coped joint is easier to produce: It doesn't require the perfectly square corner that a mitered joint needs. A coped joint is also less likely to open after a few seasons of expansion and contraction.

Although you need a miter saw for coping, the only specialty tools you need are a $10 coping saw and an assortment of blades. A 15-tooth coping-saw blade is the best all-around performer, especially for simple chair rails and baseboards. But you'll want an 18-tooth (or more) blade to negotiate the intricate cuts that crown molding requires.

When installing a new blade, make sure the teeth face forward (the same as a standard handsaw), and tighten the blade securely.

Turn your jigsaw into a supercharged coping saw

If you measure your trim in miles rather than feet, you might want to invest in the Collins Coping Foot (Collins Tool Co.; www.collins tool.com). The coping foot is a curved baseplate that substitutes for the standard, flat base found on a typical jigsaw. The manufacturer offers a coping foot to fit all commercially available jigsaws. Most install with the turn of a screw, though some saws require a shim to position the baseplate correctly.

With the coping foot in place, the saw is operated upside down, which takes a little practice but allows you to see the cutline perfectly. The curved base makes it easy to back-bevel a baseboard, but the coping foot was designed for quickly negotiating the intricate twists and turns that crown molding requires. Instructions for coping crown molding using a simple jig are included with the tool.

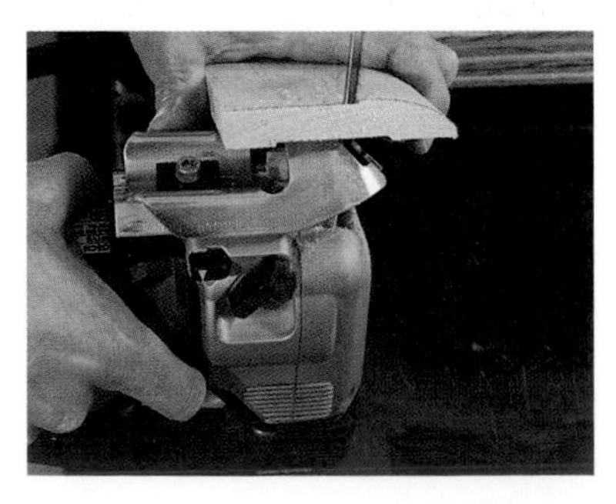

FLAT MOLDING IS STRAIGHTFORWARD

MITER CUT DETERMINES THE PROFILE. Although you can trace the profile from one piece of trim to the other, a 45° inside miter cut achieves the same purpose. An efficient carpenter chops all the profiles for a particular room at the same time, then cuts each piece of trim to length later.

USE A PENCIL TO HIGHLIGHT THE CUTLINE. To make the profile of the molding more apparent, draw the flat edge of a pencil lead across the inside edge of the miter cut.

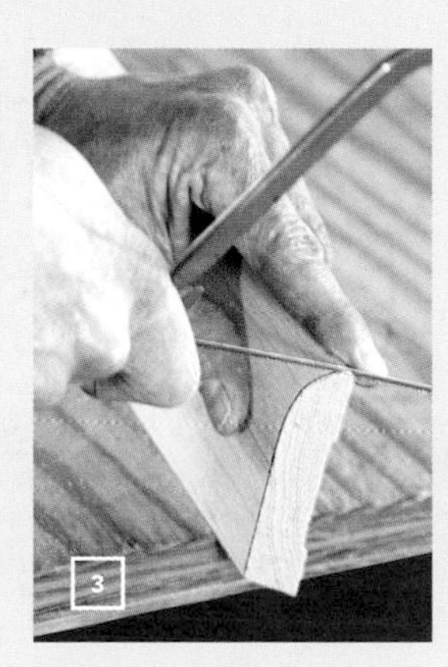

ANGLE THE CUT INWARD. Start the cut with a few gentle pull strokes until the coping saw finds its groove; then switch to long push strokes.

ANGLE THE BLADE INTO THE WORK so that the face of the cut becomes slightly proud of the back side. This slight angle is called a back bevel, or back cut.

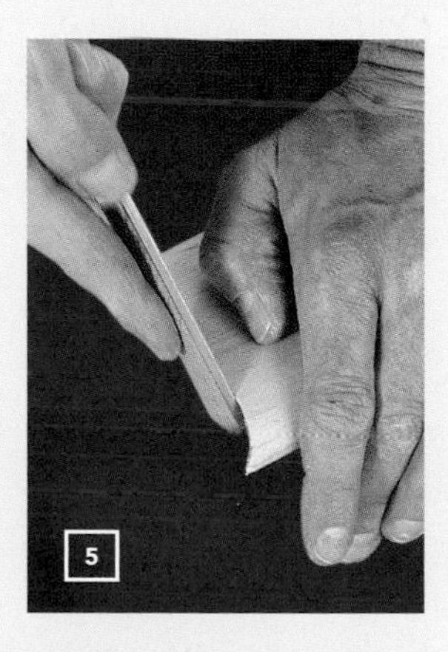

THE BACK BEVEL ALLOWS MINOR ADJUSTMENTS to be accomplished using a few passes with a wood rasp rather than a belt sander.

CROWN MOLDING TAKES PATIENCE AND A STEADY HAND

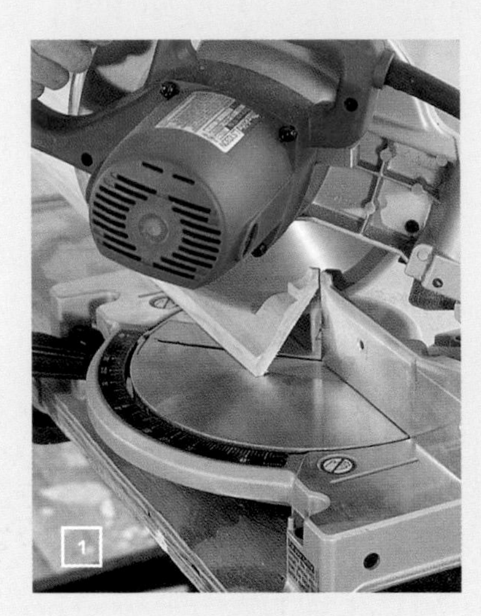

PLACE THE CROWN MOLDING UPSIDE DOWN in the miter saw and at an angle between the fence and the base. Then make a 45° cut to reveal the profile for the cope.

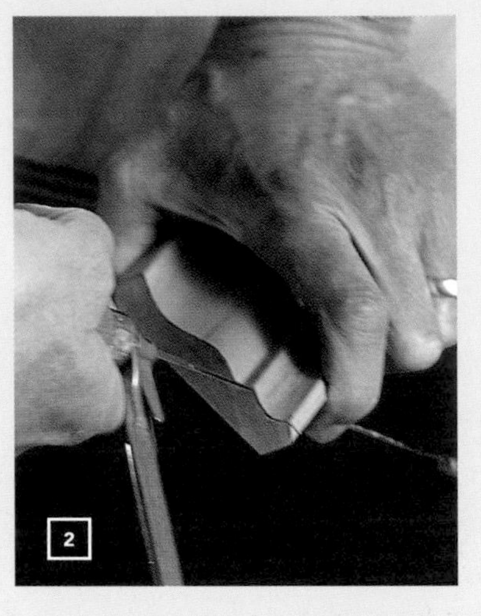

CROWN NEEDS A STEEP BACK BEVEL. Because it's installed on an angle, unlike baseboard, crown molding must be coped with a significant back bevel, or the two faces won't meet.

IT'S NOT EASY TO TURN corners when sawing at such a steep angle, so the best strategy is to cut as far as you can from one end, back the blade out, and sneak up on the cut from another direction.

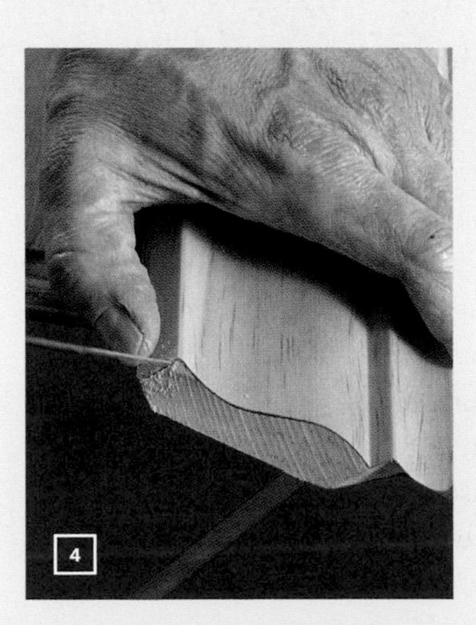

WORK INWARD FROM BOTH EDGES to ensure that the last saw stroke separates the meaty center of the molding rather than the fragile outer edge.

CLOSE WON'T DO. If the cope doesn't fit perfectly, use a pencil to mark the high spots, which you can remove easily with a rasp or some sandpaper.

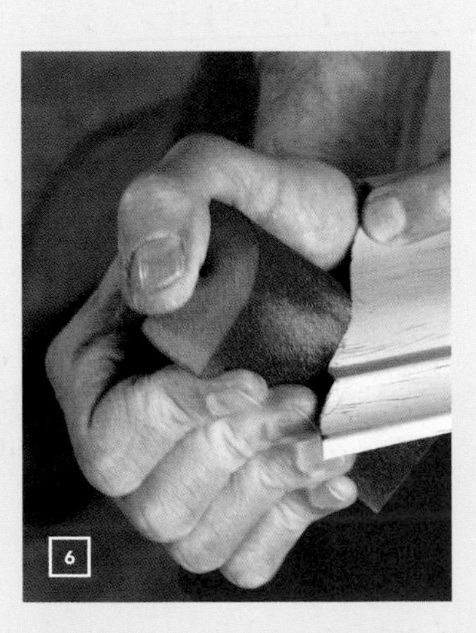

FINE-TUNE THE CURVES. A medium-grit sanding sponge is particularly effective for shaping curved sections.

The Secret to Coping Crown Molding

BY BILL SHAW

Coping crown molding is not thought of as a science, but more as an art. It involves a set of skills and techniques passed from master to apprentice, or less formally, from one guy on the job to another. After working with trim for a long time and talking to a great many carpenters, I have developed a good handful of crown-coping techniques, and I'd like to pass them along. Created out of need, the techniques are based partly on math, partly on common sense, and partly on learning from mistakes.

The most important concept to realize is that every crown profile is designed for a fixed wall and ceiling projection. Once you have determined the projection, you need to reproduce the measurement accurately and consistently in your miter cuts and in the installation of the coped joint. This will always remove the complications brought on by irregular corners, bad framing, and lumpy tape jobs. Cut precisely, a coped joint is forgiving and fits every time.

Transfer the ceiling projection to the saw

Every crown profile has a specific wall and ceiling projection. If the crown is cut at the same angle as the projection, the cope will fit every time [1]. To find the ceiling projection of the crown, I fit a square onto the bedding angles of a flat sample piece; the ceiling projection is indicated on the top scale [2]. Rather than use a compound-miter saw to cut crown on the flat, I've found that it's faster to create a saw-table stop that reproduces the ceiling projection, then cut upside down and backward. After measuring the projection, I rip a piece of plywood to the width of that measurement (in this case, 3⅛ in.). I use the plywood to set the position of a stop screwed to the auxiliary table [3]. With the fence in place, all cuts will be equal [4]. To check the cut, I measure the length of the miter; it should equal that profile's ceiling projection. Installed at that projection, the cope will fit.

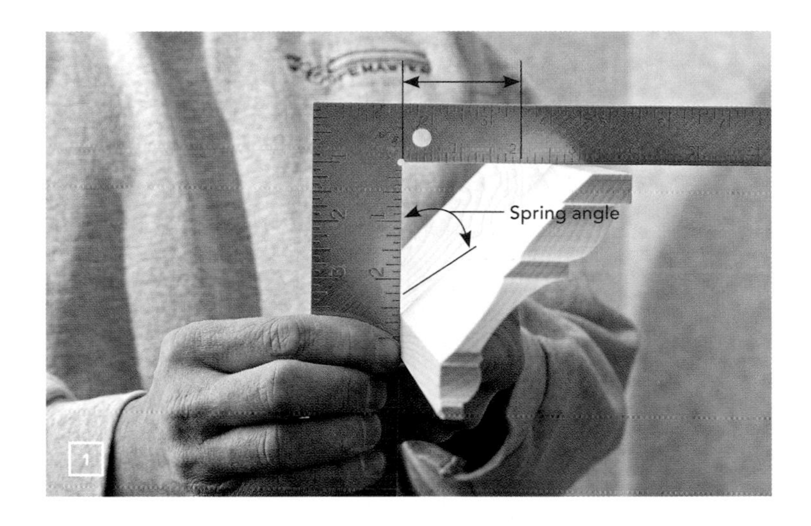

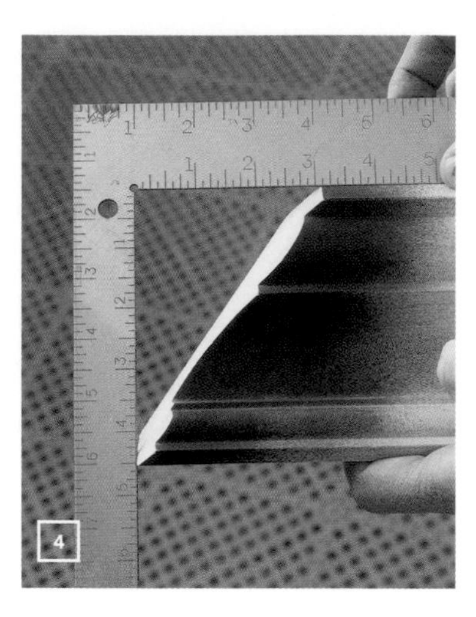

Use gauges, and take the guesswork out of the installation

Sprung crown doesn't make contact with the intersection of ceiling and wall, so measuring from the corner to mark guidelines only introduces errors that result from framing irregularities or joint-compound buildup. Instead, I use the piece I ripped to set the miter stop, glue a slice of the crown into position, and lop off the corner. With the gauge held up to the ceiling, I can mark the crown's top and bottom positions 5 6. An alternative technique comes from my colleague Dave Collins (www.collinstool.com), who cuts the wall and ceiling projections of a crown profile into a piece of plywood that holds the crown in the proper position. It's especially handy for laying out and installing outside corners. I measure the angle of the corner, subtract that angle from 180°, and divide by two to get the miter angle. Once the miters are cut, I install the piece, using the gauge to hold it at the proper projections 7. My favorite way to establish a guideline is to use a 2-ft.-long piece that's coped accurately and a 2-ft.-long square-cut piece. Fitted in the corner, the coped piece locates the square piece at the correct projections, and you can mark the top and/or bottom. The longer pieces take into account more of the wall and ceiling conditions than small gauge blocks.

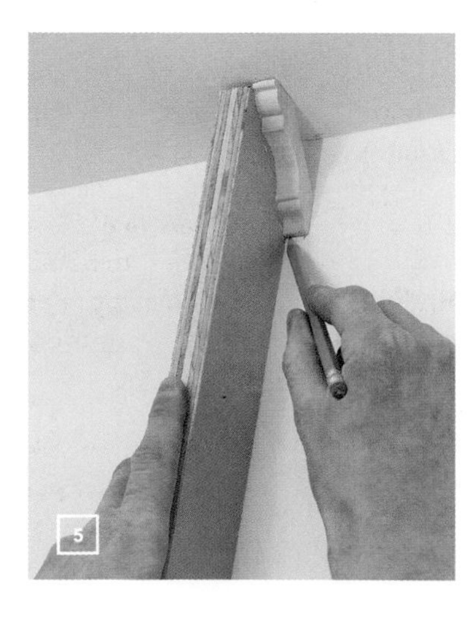

A coped joint that looks mitered

Especially when you're working with wide stain-grade stock, there are occasions when you'd like the crisp look of a mitered joint but want to keep the integrity and ease of the cope. (See "Baseboard Done Right" on pp. 49–58) Both sides of the joint need a little attention [8]. After cutting the miter, I cope away the waste but leave the point intact [9]. I test-fit the mate of the piece and scribe the point onto the opposite piece, then make a shallow cut with a fine-toothed handsaw.

CARING FOR A TRIM DELIVERY

ATTENTION TO DETAIL is an important attribute of finish carpentry. Molding quality can have as much to do with the outcome of a trim job as cutting and installation techniques. Here's the list of things I check over in a trim delivery.

- **Moisture content** Use a moisture meter to check the trim. Make sure the moisture content is within an acceptable range for your area of the country. You can call a local mill to find out the recommended moisture content for a particular species of wood.
- **Consistency** Measured with a dial caliper, the trim should not vary by more than 1/64 in. in width over its length or from piece to piece.
- **Flaws** Be on the lookout for excessive chatter, tearout, and other imperfections.
- **Relief cuts** If possible, have the supplier mill a relief into the back of the molding. The relief helps to reduce cupping, and it allows the trim to sit flatter on the wall. I prefer one big relief over two smaller cuts on base or large casings; if the stock does cup, it will rock on the area between the two relief cuts.
- **Cupping** Check profiles with a straightedge. A slight amount of cupping is normal on some pieces, especially with deep profile cuts, but anything with severe cupping should be sent back.
- **Bed angles** The crown's bed angles should fit flush in a framing square or be slightly undercut at the ceiling. A 2° to 5° undercut helps if the ceiling is going uphill or if you have to alter the spring angle of the crown.
- **Climate control** If the trim is to be acclimatized, the job site must be as close as possible to the homeowners' living conditions. Ideally, the relative humidity should be 35% to 40% at a temperature of 60°F to 70°F. Prime the stock on all sides, don't store it in the garage or basement, and if possible, use air conditioners or dehumidifiers to offset job-site moisture.

Site-Made Moldings in a Pinch

BY KIT CAMP

Our little 1920s house suffered more than a few "improvements" before we purchased it. The most egregious was the installation of cheap vinyl windows. To add insult to injury, the installers didn't bother to match the existing trim when they replaced the apron moldings under the new windows. Although we have yet to remedy the window situation, I decided I could at least install some matching trim.

Because of the age of the house, I couldn't find a stock profile to match the aprons, and I didn't want to pay to have the profile custom-milled; I needed only a couple of 8-ft. sticks. I decided to make the trim myself using my tablesaw, a few hand tools, and a technique that I've used in the past to match baseboard, door casings, and crown in a pinch. I also use aspects of this technique to make profile-specific sanding blocks for fairing scarf joints on long runs of trim.

While not a speedy process, this technique can save you a lot of money in router bits, custom shaper knives, or order minimums at the lumberyard. That said, it's difficult to reproduce some smaller, more intricate details without the help of old-fashioned molding planes, scratch stock, or custom-made scrapers, so take a hard look at the molding you need to duplicate before jumping in. If you are on the clock, a practical limit is around 16 ft. of trim.

Take your time at the lumberyard, and look for quartersawn stock that has straight grain to use for moldings. Poplar works well for painted interior moldings; fir and redwood are good for exterior use.

If possible, make a clean, square cut in a scrap of the molding to be copied, and use this scrap to trace the profile to the end grain of the new stock. Old trim often has many layers of paint, which must be scraped away to reveal the original profile. If you can't use an actual piece, trace the profile onto a 3×5 index card. The profile then can be transferred to the blank stock. You can do this without removing the old trim; make a thin cut in the trim using a thin-kerf pull saw, slide the index card into the kerf, and trace the profile.

WHEN ALL YOU NEED IS A FEW FEET OF MILLWORK to match existing trim, look to the tablesaw, a block plane, and some sanding blocks.

1 SCRIBE THE PROFILE TO THE STOCK. The first step is to mill the stock to the same dimensions as the molding to be copied. If it's possible, use a square-cut piece of the original molding to trace the profile onto the end of the stock. This cutoff later can be used to set the tablesaw blade. If the molding can't be removed, trace the profile onto an index card, then transfer this to the stock. A fine-point marker offers a thin, clean line in most cases, but if the wood is dark, consider a white-colored pencil.

2 START WITH RIPS ON THE TABLESAW. With the profile marked clearly, make a series of overlapping rips on the tablesaw, usually moving the fence ⅛ in. or less each time. It helps to leave a flat section of stock on each edge and in the middle so that the wood runs across the saw table evenly. Play around with the saw's bevel angle and its depth of cut to get as close as possible to the desired profile line. A full-kerf blade with a rip-grind (flat bottom) tooth pattern will remove more per pass and save you work later.

3 DIAL IN THE DETAILS WITH A BLOCK PLANE. A sharp block plane is the hero of this technique, but a shoulder plane (see the right photo above), a small rabbet plane (Stanley® #75 or equivalent), and an assortment of curved-sole and miniature planes are also helpful. If you don't have a block plane, rough-grit sandpaper will do the job. Whether planing or sanding, the goal is to work the length of the piece evenly, taking long, light passes from one end to the other and being careful to keep the details crisp.

4 SMOOTH THE PROFILE WITH SANDPAPER. Custom-made wood sanding blocks and store-bought sanding backers (see the top right photo) help to achieve fair curves and crisp corners.

For painted molding, start with 80-grit sandpaper, and finish with 120-grit paper. For stain-grade moldings, continue to sand up to at least 150-grit; 220-grit is even better. Make the blocks as long as possible for the most consistent results, and take even strokes that run from one end of the workpiece to the other.

5 CUT BEVELS TO MATCH THE ORIGINAL MOLDING. If the molding has a beveled back side, use the original piece to set the tablesaw to the appropriate bevel angle (see the photo on p. 81), and rip the molding to match (see the left photo above). In some cases, this bevel will be steeper than 45° and might need some work with a block plane. Once ripped, compare the new molding to the original, and do any final detailing for a perfect match.

PART 3

Trimming Windows and Doors

Trimming Windows

BY JIM BLODGETT

We constantly look through windows. A well-trimmed window enhances the view the same way a picture frame enhances the art it surrounds. As with a picture frame, window trim should stand up to critical scrutiny but not draw your eye. Doing a good job trimming windows is a matter of following an efficient sequence, understanding what has to be done when, and slowing down enough to do your best work.

I usually trim windows in one of two ways. They either have stools, the horizontal piece at the window's bottom where the cat sits (see the drawing on p. 86), or they don't and are picture-framed with casing. In either case, since the energy crisis of the 1970s brought on the demand for thicker insulation, most houses I work on are framed with 2×6 studs. Windows, though, are generally still configured for 2×4 walls, so I have to extend the jamb to bring the window flush with the drywall.

I usually trim the widest, tallest windows on a job first. This way, if I cut something too short on the larger windows, I might be able to recut and use the trim on a smaller window. I also identify the most visible windows on the job, usually the kitchen, living room, entry, and master bedroom, and try to use the best-looking stock in those high-visibility areas.

First, mark all the details on a story pole

Before cutting stock, I trim any drywall that extends past the framing and remove any blobs of taping mud. Drywall dust is extremely abrasive and will quickly scratch trim, particularly prefinished material. Because of this risk, I take a minute to sweep the room clean after I prepare the openings.

Then, assuming that I'm trimming more than a handful of windows, I make up story poles (see the photo and drawing on p. 87). Story poles are lumber or plywood scraps with all the relevant dimensions for a job marked on them full scale.

I refer to story poles throughout the job for measurements. Story poles reduce the possibility of errors in measurement that can happen when reading a rule or tape. I use two story poles for each size of window: one each for height and width.

After labeling the marks and double-checking all reveals and measurements, I cut the poles to length at their widest marks. A word of caution: Sometimes the reveal between the bottom jamb extension or stool and the window will differ from the other three reveals. Casement windows often have cranks at the bottom, and you need to be certain that you have left the space for them to turn.

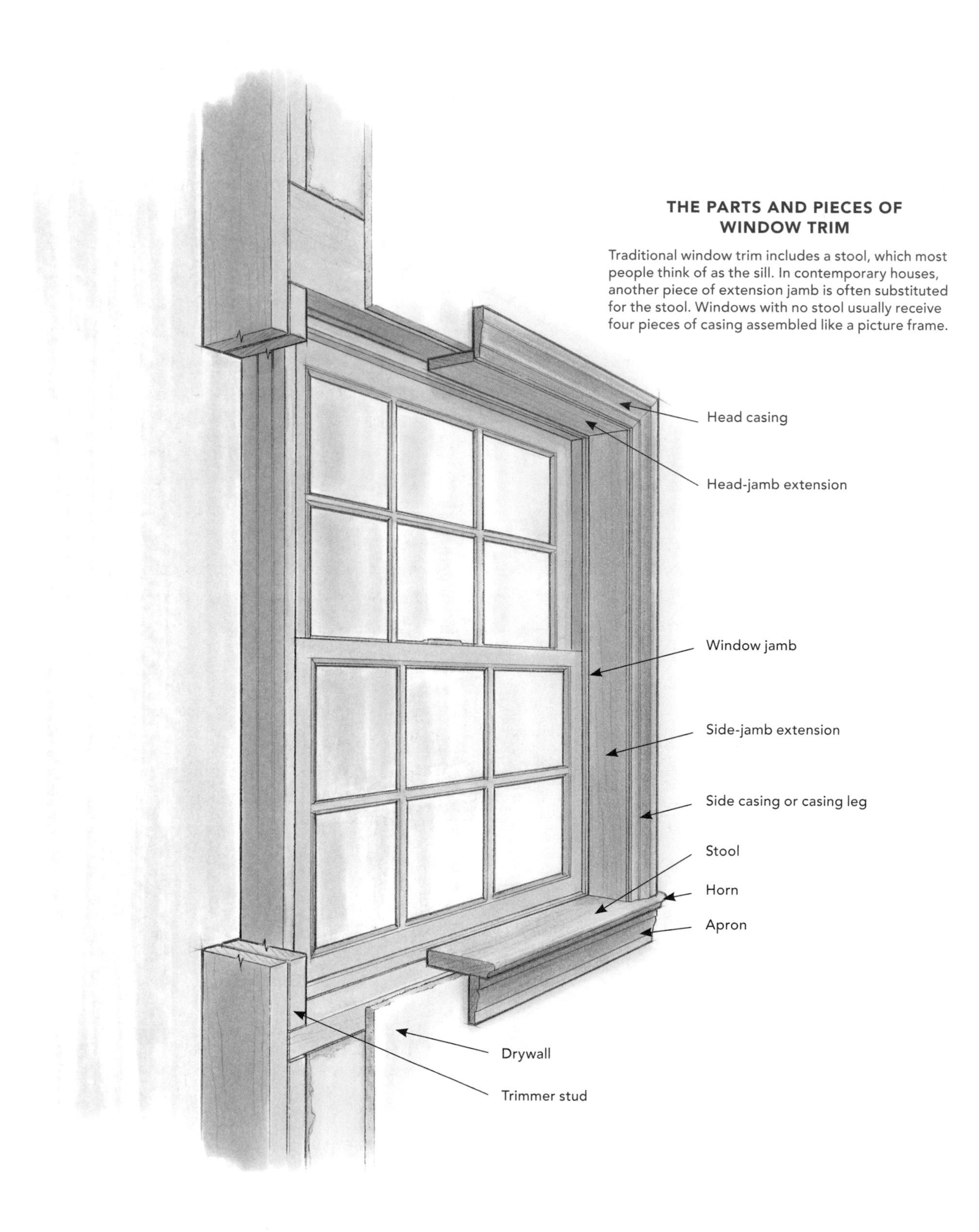

THE PARTS AND PIECES OF WINDOW TRIM

Traditional window trim includes a stool, which most people think of as the sill. In contemporary houses, another piece of extension jamb is often substituted for the stool. Windows with no stool usually receive four pieces of casing assembled like a picture frame.

To get trim straight, you might have to install crooked jamb extensions

I start by ripping jamb extensions, usually nominal 1-in. stock, slightly wide on a table saw. This width leaves some meat for trimming. I set my blade at 2° or 3° to slightly bevel the edge of the jamb extension that meets the window. The slight point of this bevel is on the visible side of the jamb extension and helps to ensure a nice fit.

When possible, I cut all four jamb extensions for a window consecutively from a single board. This technique keeps the boards color-matched, and the grain usually meets at all but the last corner, where the opposite ends of the board meet up. I think about how this window will most often be viewed and put the last joint where it seems least conspicuous. I usually cut the top and bottom jamb extensions the same length as the window width plus reveals, and cut the side jamb extensions to lap the top and bottom ones.

But I'm getting ahead of myself. Before cutting extensions to their final lengths, I rip them to their exact widths. Because it's likely that I'll be custom-tapering each piece, though, I have to cut the long stock close to its final length. I cut all the pieces 2 in. oversize and number their backs sequentially.

So that the subsequent casing joints will lie flat and fit well, it's critical that the jamb extensions end up in plane with the surface of the wall, particularly at the corners. Because windows rarely seem to be installed exactly parallel to drywall, jamb extensions usually need to taper to come up flush with the drywall. On new construction, it's not unusual to have to taper the jamb extensions as much as 1/8 in. There's no telling what you'll find on a remodel. The alternative is bashing the drywall even with the jambs, an inelegant solution that at best gets the two only close to flush.

I find the widths of the jamb extensions at the windows' corners by measuring to a straightedge held against the finished wall (see the top left photo

One vertical and one horizontal story pole per window speeds accurate production by providing a ready benchmark for cutting all the pieces of trim.

STORY POLES CONTAIN ALL THE INFORMATION NEEDED to cut the stock to length. This includes the width of the window, the reveals between the window and the jamb extensions, and between the jamb extensions and the casing; the width of the casing; and finally, the distance the stool, if any, extends past the casing.

EACH CORNER IS MEASURED SO THAT JAMB EXTENSIONS CAN BE CUT EXACTLY. Even on new construction, it's common for the window to be considerably out of plane with the wall.

SET SHIMS BEFORE INSTALLING THE JAMB EXTENSIONS. The reveal added to the thickness of the jamb extensions gives the author the distance that the shims must be from the edge of the window jamb.

above). Then I taper the jamb extension as needed on the table saw (see the photo and drawing on the facing page).

I test-fit the jamb extensions once they're tapered, then nail them into a boxlike unit. Before setting this unit, I nail cedar shingles to the studs; the shingles shim the jamb extensions to leave the proper reveal on the window jamb (see the top right photo). I'll typically shim every 2 ft. or so.

Next, I slide the assembled jamb extensions into place. I nail them just below the shims, using 6d or 8d finish nails. This step allows me to fine-tune the shims if necessary. When I like the reveal, I cut the shims to length using a sharp knife. Now is the time to stuff the space between the jamb extension and the framing with insulation or minimally expanding foam.

INSTALLING THE JAMB EXTENSIONS AS A UNIT. After tapering the pieces individually, the author cuts them to their final length and assembles them into a unit with the stool.

Window stools are the traditional lower-jamb extension

Windows with a stool in place of the bottom jamb extension are only a little more complicated. I start by cutting the stool to length, based on marks on the story pole. If I plan to return the ends of the stool to hide the end grain (see the drawings on p. 92), I cut the stock several inches too long. Some carpenters say that hidden end grain marks a professional trim job; I say it depends on the nature of the job. For example, I would not return the horns on Craftsman-style window stools because exposed end grain is a characteristic of the style. No design rule is hard and fast, not even one seemingly as basic as no end grain showing. After cutting the stool, I temporarily support it at its finished height on shims to lay out the notch that will bring the stool in contact with the window (see the top left photo on p. 90).

With the stool centered on the window and tight against the drywall, I measure the distance between it and the window next to each trimmer stud. (It's common for measurements to differ at each end.) I square out the appropriate distance on each end of the stool, marking where I will crosscut the notch that fits the stool between the trimmers.

Without moving the stool, I scribe it to the wall (see the top right photo on p. 90). Using the scribe line as my starting point, I mark on the stool the measurement between the stool and the window. I mark the stool in two spots: at the end of the stool and at the crosscut mark. Connecting the points marks the rip cuts. There is often a 1⁄16-in. or 1⁄8-in. gap between the wall and the end of the uncut stool. Laying out the stool in this manner should account for these gaps and fit the stool tightly to the wall and to the window.

Because the rip cut is seldom parallel to the front of the stool, I cut it quickly with a sharp handsaw (see the bottom photo on p. 90). However, I usually make the notch's crosscut with a chopsaw. The remaining part of the stool that extends onto the wall is called the horn. I test-fit the stool and make any minor adjustments to fit it tightly to the window and to the walls with a plane or a rasp. Although the bottom of the casing leg covers much of the horn, its ends will show, so I take a little extra time to get a good fit against the wall.

When I like the fit, I mark where the jamb extensions will meet the stool. A couple of 6d finish nails through the bottom of the stool in the ends of the jamb extensions holds them together. The top jamb extension is nailed to the legs, and I slide the

TAPERING JAMB EXTENSIONS WITH A TABLE SAW

To make up for framing discrepancies, jamb extensions usually need to be custom-tapered to end up flush with the wall. The author marks the jamb width on the scrap areas. Then he affixes a wider straightedge, say a 4-in. ripping of plywood, to the jamb extension with its edge 1 in. in from the marks. With the rip fence set 1 in. over the straightedge's width, passing this assembly through the saw rips the taper exactly.

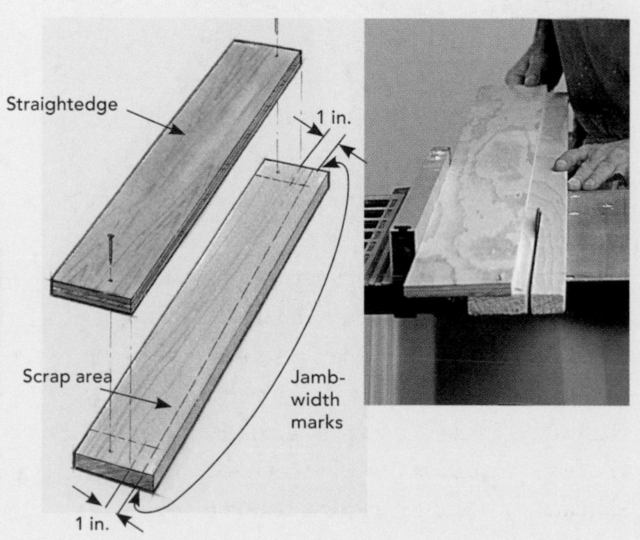

SHIMS SUPPORT THE STOOL FOR LAYOUT. The author marks the notch cut on the stool, then measures between it and the window to lay out the rip cut for the horn.

SCRIBING THE STOOL TO THE WALL. Because the wall is rarely parallel to the stool, this mark, not the edge of the stool, is the baseline for laying out the horn's rip cut.

NOTCHING THE STOOL. Although he uses a miter saw to make the crosscut, the author prefers a handsaw for the rip.

assembly into the window opening (see the bottom photo on page 88).

Sometimes I hide the end grain of the horn with returns, where the end of the horn is mitered and a matching mitered piece of stool is glued in place. I begin by leaving the stool a few inches long on both ends. After notching the horn, I miter-cut the returns from it; then I mark and miter-cut the stool to its overall length. Mitering the return pieces from the horn scrap lets me wrap the grain around the corner.

Picture frame–style casing calls for four perfect miters

With the jamb extensions in place, the window is ready for casing. As with the jamb extensions, I try to cut all four pieces of casing consecutively from a single board.

Getting all four miters tight can be tricky. I've seen the air turn blue and the kindling pile grow as a carpenter tried to fit a board that was mitered at both ends. If the jamb extensions are slightly proud of the wall, either because they were cut wrong or because nailing the casing to the wall pushes the drywall tighter to the studs, the casing cants slightly backward. This causes the pieces to hinge on the backside of the miter, opening the face of the joint. I avoid this problem by slightly carving out the ends of the miters with a knife (see the middle left photo below).

Taking the measurements from my story pole, I miter both ends of the four pieces of casing. I install the head piece, then the two legs and finally the bottom piece, changing mitered angles slightly with the saw or block plane. I back-cut as needed and work my way around the opening.

I use just enough nails or brads to hold the casing in place until the joints look good. I first nail the casings to the jamb extensions, then glue and nail the miters (preferably from the top and bottom), using 3d finish nails for both (see the below right photo). I finally nail the casing to the wall using 6d finish nails.

When picture-framing windows, I sometimes set up the miter saw remotely and cut and number all the casing pieces in a house before ever nailing any up. This technique can speed things considerably by minimizing trips back and forth to the saw.

Casing windows that have stools

As with picture-frame windows, I try to cut all the pieces of casing consecutively from the same board. On windows that have a stool, however, taking only the two legs and the head from the same board is okay.

A GAP BETWEEN THE CASING LEG AND THE STOOL STICKS OUT LIKE A SORE THUMB. A casing scrap is used to check if the stool is square to the jamb. The author trims the square windows first, then comes back to custom-cut the casing for the out-of-square windows.

YOU DON'T SEE IT, AND IT JUST GETS IN THE WAY. Removing some stock behind the face of a miter, or back-cutting, makes getting a perfect-looking joint easier.

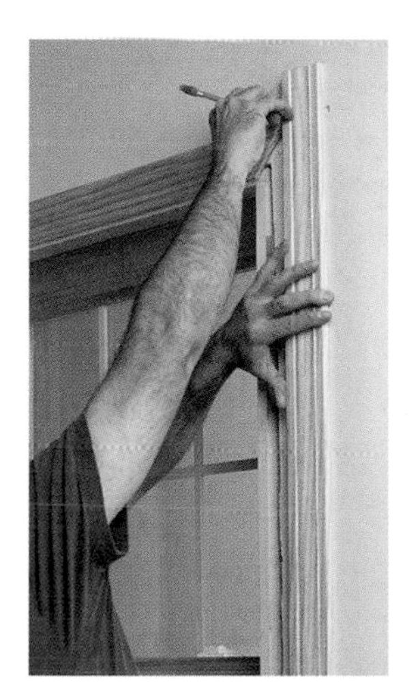

YES, THE CASING LEG IS UPSIDE DOWN. With the casing leg's point on the stool, the author marks its height. On windows that are trimmed with stool, marking the casing legs in place is more accurate than using the story pole.

TACKING THE TOP OF THE MITER LEAVES A NAIL HOLE THAT NEEDS NO FILLING. This nail does little to tighten the miter, but rather helps to keep the head and leg in plane when they're nailed to the wall.

RETURNING THE STOOL

SOMETIMES I HIDE THE END GRAIN of the stool's with returns, where the end of the horn is mitered and a matching mitered piece of stool is glued in place. I begin by leaving the stool a few inches long on both ends [1]. After notching the horn, I carefully miter the return pieces from it, and then mark and miter-cut the stool to its overall length [2]. Finally, I glue the return to the horn so that the grain wraps around the corner [3].

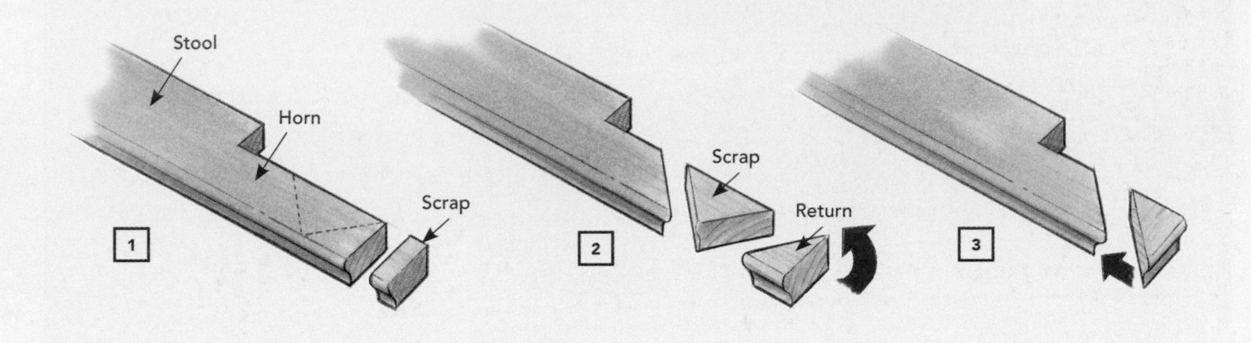

The bottom piece of casing, or apron, doesn't touch the legs. It runs underneath the stool, so grain differences aren't as apparent.

I cut and install the head casings first, again referring to the story pole for dimensions. I then miter the legs, one left and one right for each window, leaving them an inch long, and stack a pair at each window. As I visit each window, I hold a square-cut casing scrap in place to check if a square-cut leg will fit against the stool (see the top left photo on the facing page).

I mark the casing legs' length (see the bottom left photo on p. 91). This step shows me its length, and I cut and test-fit the leg, checking for any discrepancies in miter or length. If everything looks good, I back-cut the miter, spread a little carpenter's glue on the miter, and nail the leg in place.

I don't use story poles to lay out the casing legs of windows that receive a stool because in this case, marking the legs in place is more accurate. This is particularly true if the window is slightly out of square and if I have to cut the leg's bottom at an angle.

For out-of-square windows, I use a bevel gauge to find the angle at which to cut the bottom of the casing legs. I cut these legs a bit long and test-fit each angle before cutting the leg to finished length. I nail up through the stools into the bottom of the casing legs to tighten those joints.

Finally, I install the aprons. For material with a profile, say a colonial casing, I usually cut mitered returns on the ends of the aprons so that end grain doesn't show (see the photos at right on the facing page). This step is similar to cutting returns for stools (see the drawing above).

Because of their small size, cutting these returns can be trickier than cutting stool returns. When I complete the cut, I release the trigger and wait for the blade to stop spinning before raising it. Sometimes the return shoots out the back of the saw, but it usually tips over harmlessly next to the blade. I then cut the required miters on the ends of the

THE FINAL ACT OF CARPENTRY WHEN TRIMMING A WINDOW IS SETTING THE APRON. Its ends should line up with the edge of the casing legs, and it should fit snugly to the stool.

apron and glue the returns in place. Pinch clamps (see the photo on the facing page; The Peck Tool Co.; www.pecktool.com) are a great way to hold returns in place until the glue sets.

Craftsman-style trim is usually butt-joined, so it goes a bit quicker than mitered trim. I cut the head so that it will overhang the legs by ½ in. or so, depending on the customer's wishes. And I install the legs first so that eyeing a symmetrical overhang is easier. Of course, on Craftsman-style trim I don't use returns. Whatever style aprons I use, I usually cut them the same length as the width of the casing, leaving the same reveals above and below the stool. I spread the aprons around the room and nail as many as I can at once. Often the wall right below a window is not perfectly flat, so I use 8d nails to suck the apron tight against the wall.

THREE WAYS TO CLAMP A RETURN

RETURNS HIDE END GRAIN. Apron returns are so small that nailing can split them and moisture in glue can warp them and ruin the fit. Clamping them until the glue sets is a good idea.

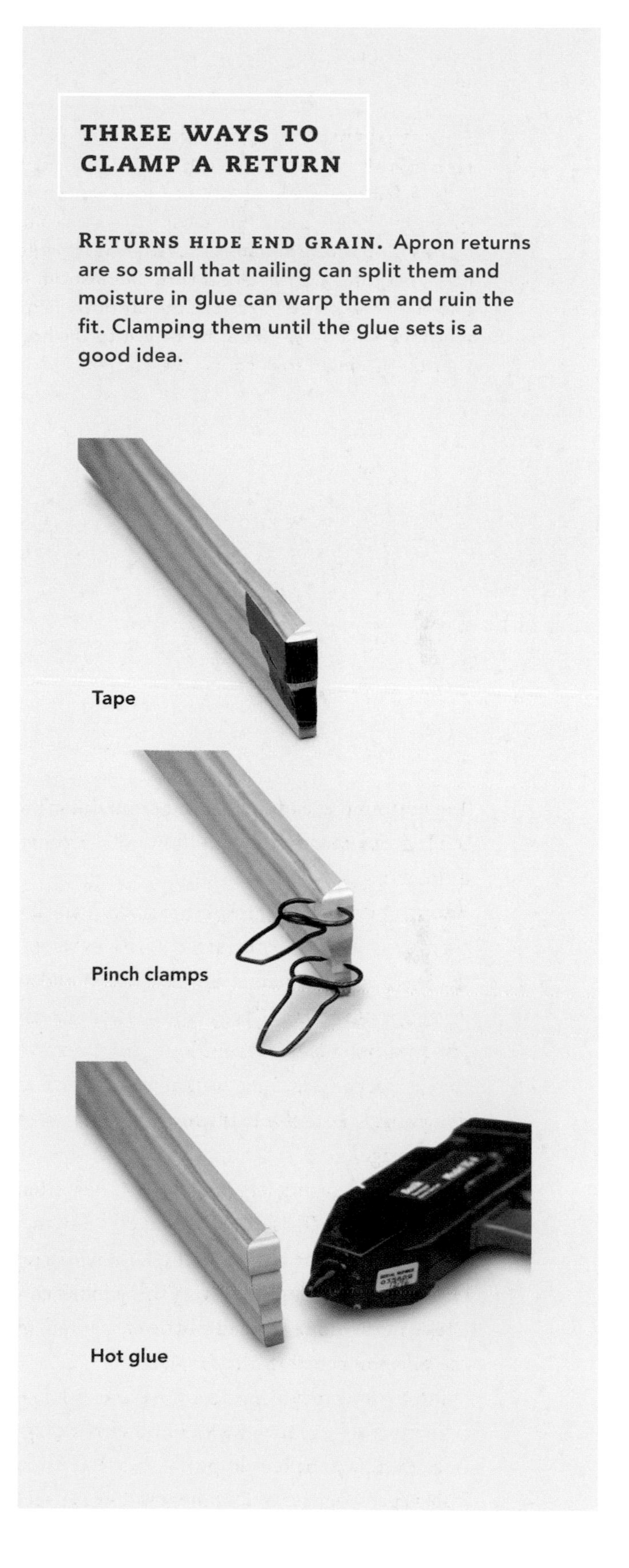

The Only Way to Trim Exterior Windows

BY MIKE VACIRCA

I can no longer call myself a trim carpenter because these days I do a little bit of everything. I was a trim carpenter for most of the past 15 years, though, and a boatbuilder before that, so I feel comfortable saying that I know a thing or two about trimming out a window, and about how water affects wood.

Every time I dropped a piece of window trim from scaffolding 30 ft. in the air, I found myself remembering another career past, the days I spent working in a cabinet shop, where work is easy to control and weather isn't a concern. Finally, I came to my senses when I was presented with 27 windows to trim for one house. That job helped me to develop a method for installing exterior trim that's easier on my body, that is safer and faster, and that also yields more durable results. To make this process as efficient as possible, I even prime, putty, and paint the casings before installation.

Cutlists and stations add efficiency

When I worked in the cabinet shop, I organized my projects with cutlists and made shop drawings that showed how everything was going to be built before a saw ever touched wood. The process I created for assembling window trim in the shop starts with that premise. As with the cabinet work, I organize the shop into efficient workstations to build the trim. I want the workflow to move so that the pieces are cut and the assembly happens in such a way that a few folks can work at the same time and not get in each other's way. The process, however, begins on site.

When the windows arrive, I grab a tape measure, get a notebook, and make a list of all the windows. I measure each to determine the finished height and width so that I can make a cutlist. I add ⅛ in. to each dimension to allow for caulk and to be sure the casing units will install easily by just slipping over the windows. Finally, I break everything into a formula (see the sidebar on p. 96). To keep track of which window is which, I use the lettered labels attached to them at the factory. Once the casing unit is built, I mark the respective letter on its back.

I install painted trim, so I build everything with primed stock. I cut all the pieces first, then assemble each unit with biscuits, glue, and screws to ensure that joints won't open over time. Once all the units are built, and the nail and screw holes are filled, I prime any exposed wood and give everything a first coat of finish paint.

Get the proportions right

The windows on the project featured here are a mixture of Marvin® (www.marvin.com) double-hung,

casement, and fixed units. All of the windows have aluminum-clad exteriors and primed wood interiors, and they are installed with a nailing flange. The nailing flange is set 1 in. back from the face of the unit, which in turn ends up 1 in. proud of the unsided building.

Because one of the functions of the trim is to protect the window, I like to use at least 5/4-in. stock. Likewise, if there is going to be a reveal between the two components in this assembly, I prefer that the trim be proud of the window. Leaving the window proud of the trim looks cheap.

Getting the trim width correct can be a bit trickier. Stock dimensions (3½-in. and 5½-in. boards) rarely create pleasing proportions. The right width typically depends on the size, the shape, and the style of the house and windows.

The windows on this house called for a Craftsman-style trim design, which I created using various reveals.

START BY MEASURING EVERY WINDOW

Most windows come marked with dimensions, but don't rely on those numbers for your trim. Measure the height and width of the finished face of the actual window. Add ⅛ in. to each dimension to allow room for caulk and a smooth installation. Finally, add the desired reveals to the lengths of the head and sill, and you're ready to cut.

FORMULAS FEED THE CUTLIST

Whether you're trimming one window or a whole houseful, these simple formulas make compiling your cutlist a piece of cake.

Head and Sill

Width + ⅛ in. + 2x (casing width) + 2x (desired reveal on each side)

Legs

Height + ⅛ in.

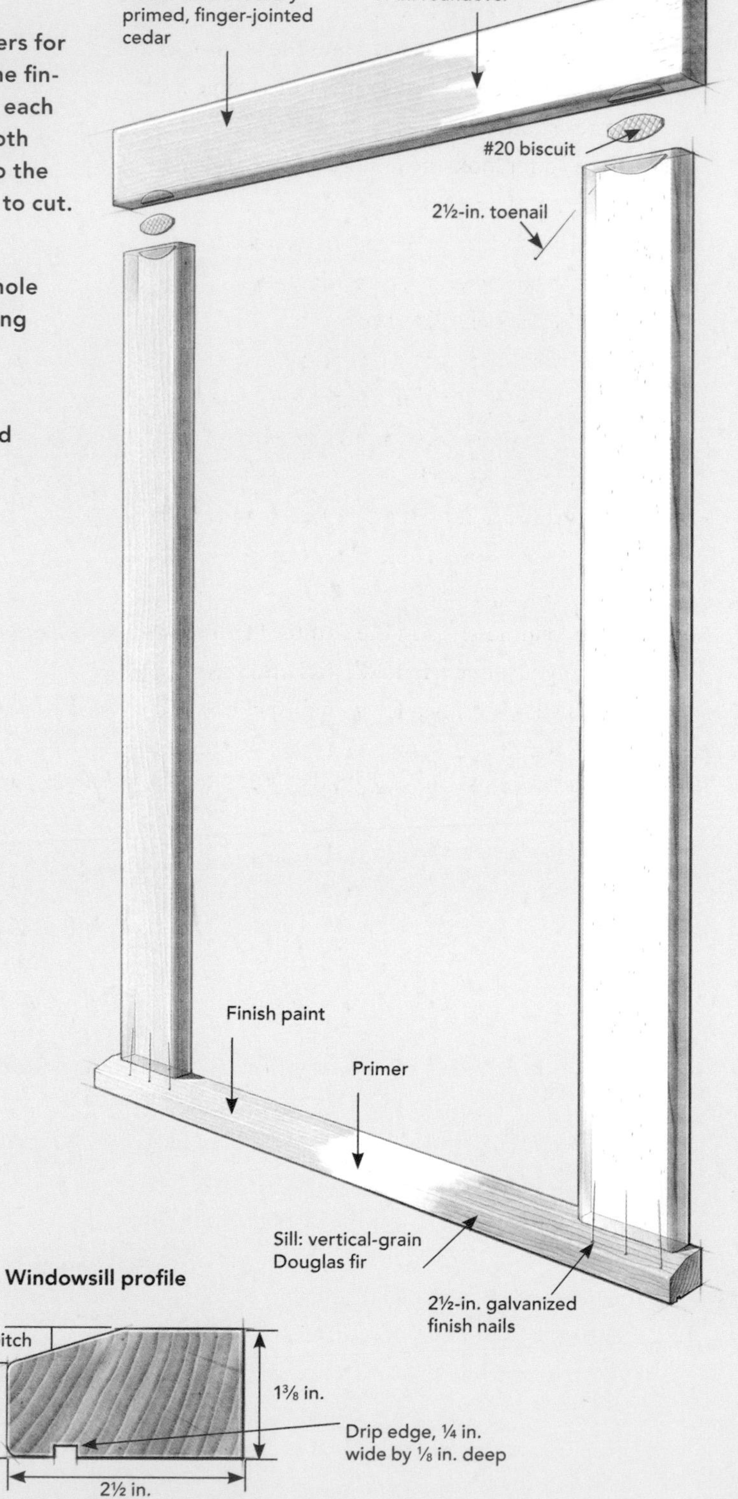

By bumping the head casing up to 1½ in. thick and by having it extend past the legs by ½ in. on each side, I was able to do two things. First, the thicker head casing provided additional protection to the window. Second, the ¼-in. reveal between the head and legs added a subtle shadowline to create an interesting look. Both the head and the leg stocks are factory-primed finger-jointed cedar. It's stable and weather resistant, and it takes paint well. I eased the edges for a softer look and a more durable finish.

I milled the sill from durable Douglas fir. The sill's shape is all about function. The top is pitched to shed water, and the bottom is kerfed to prevent water from wicking up behind the trim.

I made the sill run past the casing by ¼ in. on each side as well. I plugged the head and the sill reveals into my formula, so I was able to cut everything first and assemble the pieces later.

I've used this process on four projects now with four different window-trim styles, and the results have been uniformly awesome. The trim units go up quickly and painlessly, and the painter I work with loves making just one trip to fill nail and screw holes and to caulk, and one trip for a final (second) coat. I don't have to worry as much about the weather, and my body is thankful for the additional rest.

MILL THE SILL

USE A STABLE MATERIAL like clear vertical-grain Douglas fir (shown here). Shape the stock, add a drip edge, then prime with a high-quality exterior primer.

STRAIGHT STOCK GETS AN ANGLED RIP. Start with an appropriate length of 2½-in. by 1¾-in. stock. Rip a 15° bevel to create the sill's pitched face. Lower the blade and reset the fence. Then kerf the bottom of the sill to create a drip edge.

PLANE THE FACE SMOOTH. Use a power planer to remove saw marks and to dimension the sill's flat top to equal the depth of the window's cladding.

KEEP THE CASING SQUARE DURING ASSEMBLY

BISCUIT THE LEGS TO THE HEAD, and glue and nail the sill to the legs. Use an exterior glue such as Titebond® III (www.titebond.com). Dry-fit the pieces before assembly. One bar clamp is enough to hold things together as the glue dries. Use a framing square to check the assembly throughout the process.

SLOT THE LEGS AND THE HEAD. When the head and legs are different thicknesses, register the fence to the back of the stock. Register off the face when joining heads and legs of the same thickness.

TOENAIL THE CORNERS. Place a toenail in the inside edge of the leg using 2½-in. galvanized finish nails. This adds strength to the joint while the glue dries so that it doesn't twist during the rest of the assembly.

NAIL THE SILL TO THE LEGS. Attach the sill to the legs with glue and 2½-in. galvanized finish nails. Leave the unit clamped for an hour while the glue sets up. Prime the end grain, then fill nail holes with a solvent-based wood filler like PL FI:X Solvent Wood Filler (www.stickwithpl.com). Complete the casing with one or two coats of finish paint.

NAIL THE CASING; SCREW THE SILL

The unit slips over the window, then is attached to the house with finish nails and screws. The legs and head are caulked between the casing and the window; later, the head casing is flashed into the housewrap.

SINK THE SCREWHEAD. Nail the head and legs every 9 in. using 2½-in. galvanized finish nails. Then screw the sill to the framing. Use a ⅜-in. countersink bit to sink the screwhead about ½ in. below the surface. Fill the hole with a plug made of the same material.

Perfect Miter Joints Every Time

BY JIM CHESTNUT

Ten years ago, my approach to installing casing was fairly traditional: Measure, cut, test the fit, walk back to the saw, trim the cut. Then one day I climbed onto a joint-compound bucket to test the top piece of trim on a doorway a few feet upstream of a short flight of stairs. Microseconds later I felt my feet describing an arc about my head and proceeded down the stairs like an otter wearing tennis shoes. It's amazing how such a short trip can get so rocky, so quickly; it made me realize my methods (and stepladder) needed an upgrade.

A few months later, I happened to work with James Chambers, a talented trim and cabinet builder from Old Saybrook, Conn. He cut, biscuited, and clamped his casings together on the floor, then installed them as a complete unit. His miter joints were perfect, and the installation nearly flawless. By the next job, I had abandoned the mud-bucket shuffle and converted to the preassembly method of casing, which has improved work quality, reduced both the level of skill and hours required for the job, and extended the useful lifetime of my lower back. Here's some of what I've learned over the years, wrestling with and pinning miles of casing.

Get set up, then cut the miters

When the material arrives, my crew and I orient it in the cutting room so that a 16-footer does not have to be spun end for end in a 14-ft.-wide room. I like to cut the miters before cutting the stock to exact length and prefer to have the outside edge of the casing against the fence when cutting them; this placement eliminates any tearout on the inside edge of the miters.

I always cut two sample 45° miters after setting up the saw station to ensure that the result is square and that all the details line up. After the necessary adjustments, someone starts cutting the miters, either lefts or rights for the trim on the sides of the doors, or door legs as we call them. This way, I don't have to change the saw angle repeatedly. By placing a strip of blue painter's tape on the miter saw's extension wings as a reference, I can cut the door miters quickly an inch or so long and cut the square ends later. Once a bunch have been cut, they can be slotted for biscuits at any time. In addition, if the cutoff from one door leg is not long enough for another leg, it may be long enough for a top (or head), which can be cut after the legs are finished.

Miter clamps and biscuits make indestructible joints

To mate casing miters together so that they stay together longer than the next heating or cooling season, I assemble casings on the workbench with biscuits and glue, then clamp them together before installing them as a complete unit. I use special miter clamps that I designed and manufacture (Clam Clamps; www.miterclamp.com). The Hartford Clamp Co. makes a similar clamp. I don't believe that spring clamps exert enough pressure to allow you to move the assembly while it's still drying; I've avoided using a pneumatic nailer to pin the biscuited joint for the same reason.

Although you can hold each piece face up on the table while slotting, I find it easier to slot from the back of the casing, especially if there's limited space and a large pile of casing. Two layers of thin plastic or laminate, cut slightly narrower than the casing at 45° and glued back to back, span the relief cut in the back of the casing and form a platform that reduces the chance that the biscuit joiner's base will

CLAIM A BIG ROOM IN THE NAME OF PROGRESS. To work the customary long lengths of casing, the author advises setting up in a fairly large space that will accommodate a miter-saw station and an assembly table made of a full sheet of melamine or laminate-covered plywood screwed to a pair of sawhorses.

SLOT ALL THE MITERS AT ONCE TO SAVE TIME. After the stock has been cut to a rough length and mitered, the mitered ends are slotted for biscuits. To speed the process, each piece is partially pulled from the pile and slotted from the back; the miter's short point is used as the biscuit joiner's reference point.

Joiner fence is aligned with miter's inside edge.

Laminate scrap bridges recess in back of casing and steadies joiner.

rock, misaligning the slots. I align the outside edge of the biscuit-joiner fence with the inside of the miter and plunge. Putting the biscuit close to the inside edge of the joint helps to prevent the inside corners from opening.

Glue up is messy, especially when I'm doing it, so I use a full 4×8 sheet of plywood covered with laminate for a glue-up table, though any sheet goods (such as melamine) work fine. (Dried yellow glue easily pops off with a taping knife.) The table's width can handle most door and window casings. Wider assemblies, such as double closets, can be clamped together on the floor in front of their destinations and nailed to the jambs immediately with the clamps still attached.

I cut a bunch of 4-in. shims tapered from zero to ⅛ in.; butter the mitered edges, slots, and biscuits; shove together the joints; and align the details. I use a lot of glue and try to avoid sliding the miters against each other because it tends to remove too much glue and could result in a starved joint. I use the thin shims to adjust the alignment of the miter faces when a head or leg has a slight twist or when the glue-up table itself is torqued. When the miter is assembled, I hold it with one hand, slide a clamp into position with the other, and clamp it home. Between the glue, biscuits, and clamps, the joints set up fast; poplar casings at 7% moisture content in an ambient temperature above 70°F can be unclamped safely in 5 or 10 minutes, as long as you're careful.

Under the business end of the table, where it almost never is knocked over, I keep an old pot half-filled with water heated on a cheap electric hot

(Continued on p. 106)

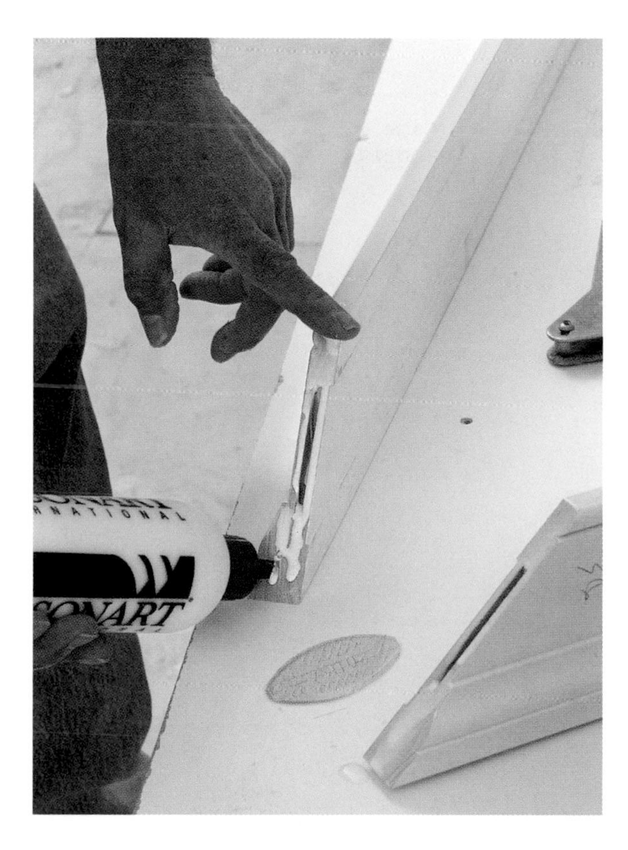

EQUAL PARTS OF GLUE AND CLAMP PRESSURE MAKE A GOOD JOINT. After spreading an even coat of glue across both halves of the miter, the author inserts the biscuit, aligns the two halves, and clamps the joint with a miter clamp. Small shims placed beneath strategic points in the casing help to align the miters.

Jim Chestnut's Clam Clamp (www.miterclamp.com)

Hartford Clamp Co.'s miter clamp

DON'T USE A TAPE; USE A STORY POLE

Printed self-stick labels (www.miterclamp.com)

Adhesive-backed tape rule (www.woodcraft.com)

TAKE AND RECORD MEASUREMENTS WITH THE SAME TOOL, THE STORY POLE. Each door is numbered and lettered: The "H" designates the hinge side, the "X" an exterior door, as in the photo above. The story pole is held on the floor, against one side of a door jamb. Here, the head jamb's bottom edge intersects the tape mark ⅜ in. above its centerline. Measurements are written as the number of sixteenths above or below the zero mark and are paired with the door's number, directly onto the pole.

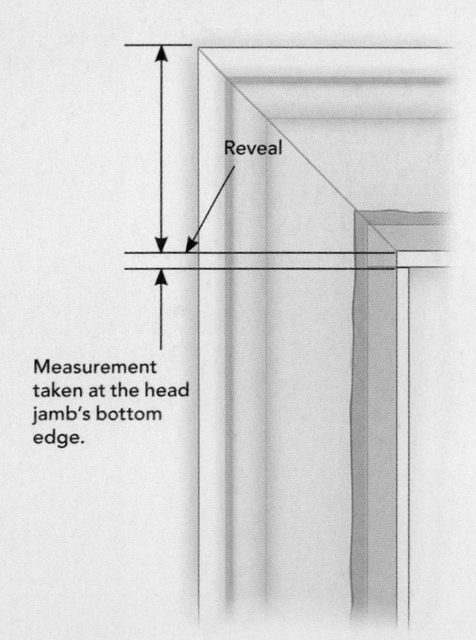

The story pole is about 4 in. longer than the longest casing leg on the job.

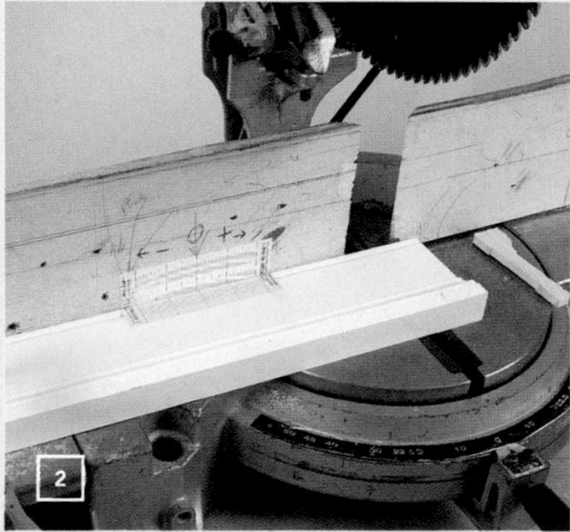

MAKE AN IDENTICAL RULE ON THE SAW. After a tape rule is applied to the chopsaw fence about 6 in. to the left of the blade, the story pole is registered exactly with the saw rule and clamped into position.

After years of using a tape to measure and mark each piece of casing on big jobs with lots of doors and windows, I stumbled onto a method that's both fast and foolproof. I use a story pole (a 7-ft. piece of casing, actually) on which I record the casing lengths; the measurements then are transferred easily to an identical marking system on the miter saw. Because I cut the miters first, each piece of casing can be measured with the mitered end registered against a stop block, then cut to length with a 90° cut.

To make a story pole, I measure from the floor to the bottom of the head jamb of a representative door; let's say it measures 81 in. I cut a piece of casing a few inches longer (say 84 in.) square on each end. Next I measure up from each end of the casing along its inside edge and carefully mark 81 in. Centered on these two marks, I adhere a piece of adhesive-backed 4-in. tape measure. I make these myself, but they're also available as lengths of adhesive-backed tape rules from Woodcraft® Supply (www.woodcraft.com); or you can replicate them on sheets of address labels using generic software.

After measuring the side jambs, I measure the head jamb, inside to inside, and write that length on the story pole as well. If I'm lucky, all similar doors' widths are identical. I then can cut all the heads of 30-in.-wide doors, for instance, at identical lengths from a fixed stop on the saw.

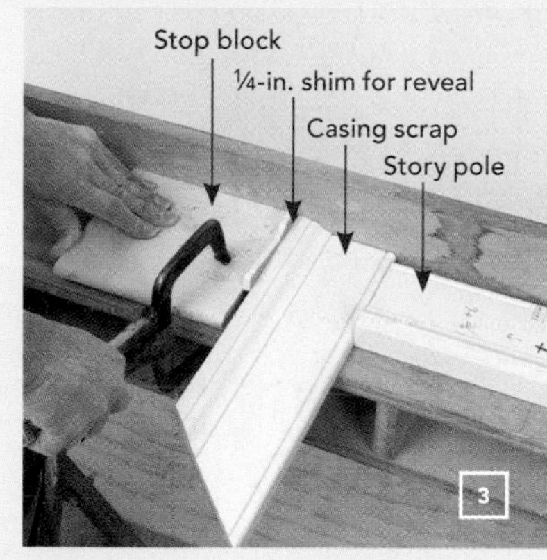

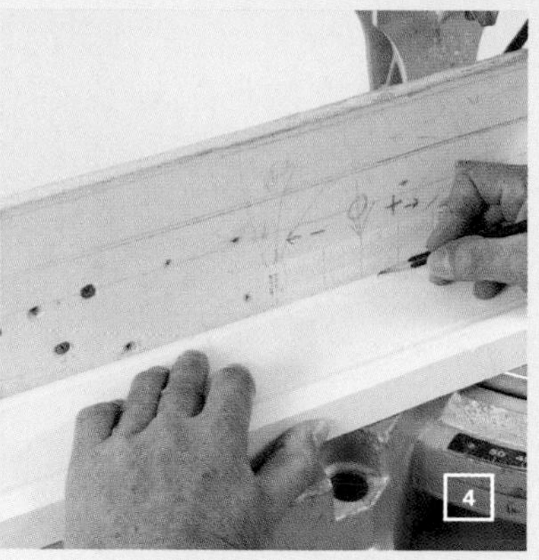

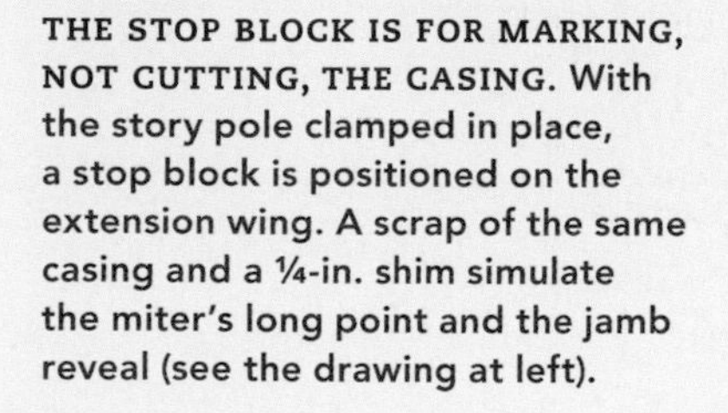

The stop block is for marking, not cutting, the casing. With the story pole clamped in place, a stop block is positioned on the extension wing. A scrap of the same casing and a ¼-in. shim simulate the miter's long point and the jamb reveal (see the drawing at left).

Mark the length, slide it over, and make the cut. With the miter already cut, the author can register a piece of the casing's miter against the stop block, transfer the correct measurement for that particular piece, then slide the piece toward the blade and make the square cut.

plate and bristling with toothbrushes. A warm, wet toothbrush cleans glue squeeze-out quickly; a quick blast of compressed air dries any residual water. I have been using this process on both paint- and stain-grade work for years without problems for the finishers, though for stain-grade or clear finish I repeat the process with a sponge and clean water after the first blast of air.

Remove any obstacles and install

Once the miters are assembled, the casings are ready to be nailed onto their respective door jambs. If the drywall is proud of the jamb, I cut away the drywall from the doorway to make sure that the casing sits tight against the jamb; I typically leave ¾ in. under the casing.

I run a thin bead of glue along the outside edge of the jambs before shooting the casings on; it's small enough that I rarely get squeeze-out at the reveal. This glue adds enormous strength and rigidity to the jambs. Although this effect is diminished tremendously with preprimed parts, I do it nonetheless. The outside edges of the casings can be nailed off later when the baseboards are installed.

When pressed by either scheduling or lack of space, I clamp and lean four casing units against the wall. I then remove the first set of clamps, clamp and lean another set, remove the next set of clamps in order, and so on, until I have as many as 16 sets complete. Then I install those with the clamps still attached, nailing the casing into the jamb only. When finished with those four, I come back and repeat the process, starting with the first set clamped. Then I take the clamps off the first batch installed and start all over again.

GOOD STOCK MAKES A DIFFERENCE

"Now the trim men will work their magic." Ever hear that phrase? That happens only when the stock is top notch. Rather than buy trim stock from the local big-box store, I pay about 20% more and get mine from a small shop that does absolutely nothing but mill trim. They use a computer-controlled molder in a humidity-controlled building and produce identical moldings from one run to the next, all with the same moisture content. This consistent stock allows you to cut just parts, rather than culling chatter, knots, cups, twists, warps, and crooks. You can join miters cut from different pieces of trim and be confident that the details will match precisely.

Once I have first-rate stock, I don't store it in garages or in unheated/un-air-conditioned areas of the house. If the house isn't ready for trim, the stock shouldn't sit at the work site. The ends of the stock absorb moisture faster than the middle, and they swell. Because the stock is no longer uniform, perfection is harder to attain.

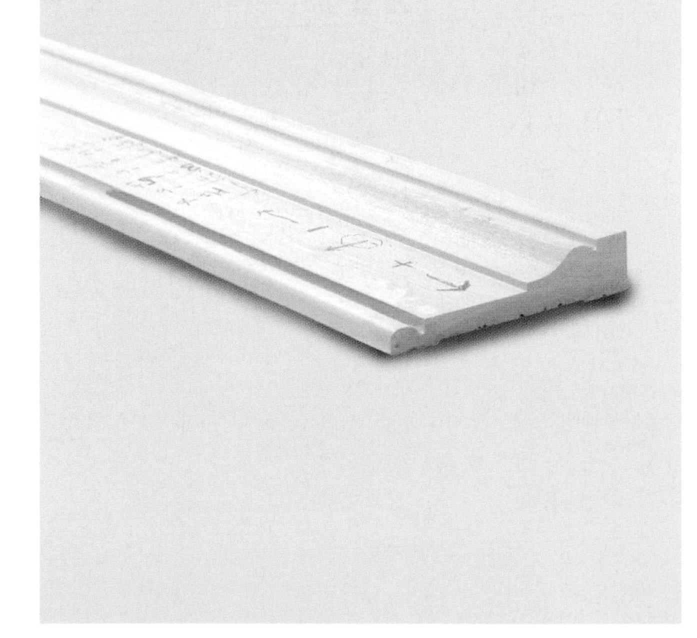

ABOVE: CLEANUP IS EASIER TO DO NOW, NOT LATER. Rather than wait for the glue to dry, the author keeps a pan of hot water and a toothbrush handy to clean up any squeeze-out. A quick blast of compressed air erases any residual water.

RIGHT: GOES UP LIKE A PICTURE FRAME. Once the glue has dried, the completed casings can be installed as a unit. Typically, the casing's inside edge is nailed to the jamb; the outside edges can be secured with longer nails when the baseboards are installed.

Trim Windows with Built-Up Casings

BY RICK ARNOLD

When I'm asked for ideas to upgrade the interior of a house, the first thing I say is, "Change the trim." The dramatic difference that built-up trim makes easily justifies the extra cost. As a carpenter, installing an interesting three-piece casing is always more fun than working with boring old clamshell or colonial casing. Because it consists of multiple pieces, built-up trim is actually much more forgiving than single-piece trim. One piece follows the window or door jamb, and a second follows the wall. Then a third piece joins the two, concealing any gaps. As my kids would say, sweet.

A few companies offer architecturally correct built-up trim arrangements (see the sidebar on p. 110). But much the same effect can be achieved with a little imagination and some stock trim from a lumberyard or a building-supply store (see "More Casing Options" on pp. 112–113). To test trim combinations, make up small sections with all the details, as I've done in the top left photo on p. 110.

Window trim starts with the stool

For built-up trim, I prefer a thicker stool with bullnose edges. To find the length of the stool, I assemble a short section of the built-up side casing. Then I set it in place near the bottom of the window, making sure to leave a $\frac{3}{16}$-in. reveal on the inside of the window jamb (see the top left photo on p. 110). I make a light pencil mark on the wall along the outside edge of the trim section (see the bottom photo on p. 110), then I repeat the process on the other side. I make the marks low enough to use as a reference later when installing the apron.

Next, I measure the distance between the pencil lines and add 3 in. The extra length allows the ends or "ears" of the stool to extend 1½ in. past the edge of the trim, rather than the ¾ in. typically used with conventional molding. The extra length also accommodates the decorative trim that will be applied to the apron.

I make sure the finished stool is deep enough for the built-up trim to land without overhanging. Ideally, the stool should extend ¾ in. to 1 in. beyond the outermost edge of the trim. To maintain the profile of the stool on all three faces, I cut 45° miters on the ends and install small return pieces that fit between the miters and the wall. I glue and nail the return pieces to the stool before nailing the stool into place (see the top right photo on p. 110).

Flat casing first

The first part of the built-up molding that I install is the flat casing. I measure from the top of the stool to the inside edge of the top jamb of the window, then add $^3/_{16}$ in. for the reveal (see the top photo on p. 111). After squaring one end of a piece of stock, I mark the length along the inside edge (in this case, the beaded edge). I cut a 45° angle using the mark as the target for the short point of the angle.

I repeat the same step for the opposite side of the window, making sure to reverse the direction of the cut. If I'm trimming a lot of windows that are the same height, I check a few to make sure they are exactly the same, then cut all the pieces at once rather than completing one window at a time.

Before nailing in the pieces, I use a biscuit joiner to cut slots into the mitered ends. Miter joints that are reinforced with biscuits are less likely to come apart over time. Taking care to keep an even reveal, I nail the side pieces into the edge of the window jamb using 1¼-in. nails (see the drawing on p. 114; photo left, p. 112).

I find the length for the top piece of casing by measuring from long point to long point on the two side pieces (see the top right photo on p. 112). After marking the top piece, I cut it just a bit long, usually by the thickness of my marks. Then I test-fit the piece, shave it if necessary, and cut the biscuit slots. To shave a hair off the miter, I place it tight against the chopsaw blade, raise the blade, turn on the saw, and bring the blade back down. By the way, I trim the wider windows first so that if I cut a top piece too short, it still can be used on a narrower window.

I glue and insert the biscuits into the side pieces with a little more glue on the mitered edges of both top and sides. I wipe the glue with my finger to ensure a thin, even coat. Next, I push the top piece into place and rock the side pieces back and forth until the miters align. If I had nailed the outside edges of the side pieces, I wouldn't be able to adjust the fit so easily. Finally, I nail the top piece into the edge of the jamb, and then I drive nails through the edge into the side pieces near the long points.

START WITH THE STOOL. After mocking up a short section of the trim to double-check the look, scribe its width on the wall to determine the length of the stool and the apron. Once cut and returned on the ends, the stool is nailed through the top into the windowsill.

The dramatic difference that built-up trim makes easily justifies the extra cost.

THREE-PIECE CASING

Combining off-the-shelf moldings creates a complex look without custom milling. The three-piece casing used in this article is from the Greek Revival series by WindsorONE (www.windsorone.com). On the following pages, you'll find designs composed of common moldings.

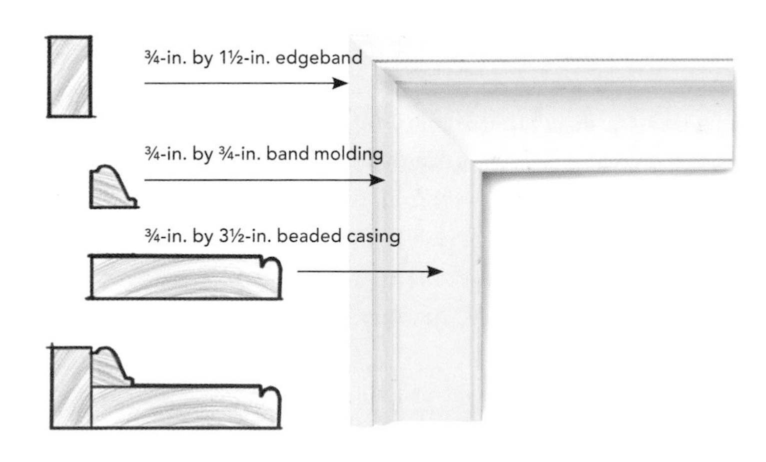

DON'T FORGET THE REVEAL. To get the length to the short point of the first side piece, measure from the stool to the inside edge of the frame, then add 3⁄16 in. for the reveal. Nail the side pieces only to the jamb at this point.

Add decorative layers

Trim pieces are added to build up the molding, beginning with the edgeband. I determine the length of the outside edgeband by measuring the outside edge of each flat side piece, from the stool to the long point (see the top left photo on p. 113). After marking that length on a piece of the edgeband, I cut the 45° angle, using the mark as the target of the short point.

The edgeband is flexible enough that it can follow all but the most-severe irregularities in the wall. So as I install each piece, I press it hard against the wall. I nail the edgeband into the edge of the flat stock only (see the top right photo on p. 113). Just as with the flat casing, I measure for the top piece of edgeband between the long points, cut it a touch long, fit it, and then glue and nail it into place.

Next, I go back and nail the flat casing to the stud framing. If the framing allows, I nail as close to the outside edge of the casing as possible so that the nail holes will be covered by the final filler trim. It's a good idea to find the edge of the framing beneath the wallboard before running any trim. At this point, I also nail up through the stool into the bottom edge of the flat casing.

To complete the built-up molding, I mark and cut the filler trim that sits just inside the outside edgeband. I use the same measurements that I took for the edgeband, only this time the measurement is to the long point of the 45° miters on the filler trim. Instead of installing the sides first and then the top, I work my way around each window (see the photo on p. 114). I make the pieces slightly long so that they spring tightly into place. I secure them to both the flat casing and the edgeband using 1¼-in. nails.

Trim the apron before it goes on

For the look I prefer, the length of the apron is the same as the distance between the first pencil marks that I made on the wall. The apron is constructed out of the same flat stock as the window casing. Just as I did with the stool, I bevel both ends of the apron and then cut, glue, and pin the small return pieces in place. Next, I cut and install the decorative trim that runs along the top of the apron. If I have a lot of windows to trim, I can work more efficiently by making all the aprons at one time.

Finally, I hold the assembled apron hard against the underside of the stool and then fasten the apron to the stud framing (see the left photo on p. 115). Also, I carefully shoot a few nails down into the apron through the top of the stool.

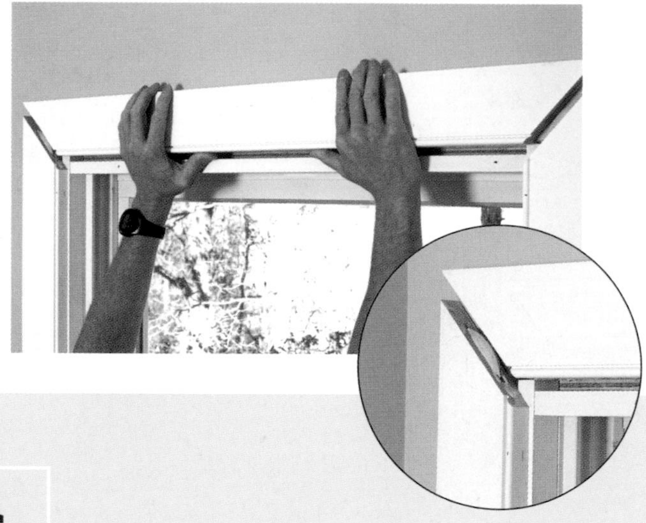

MORE CASING OPTIONS

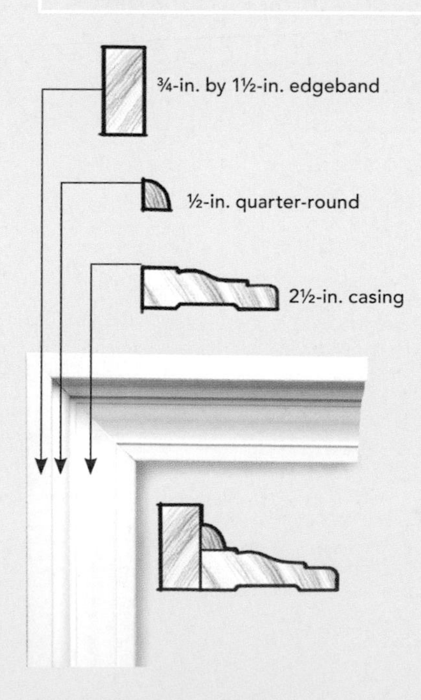

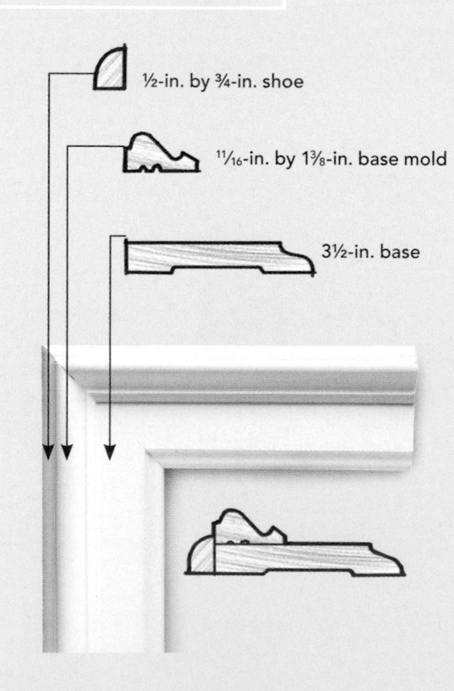

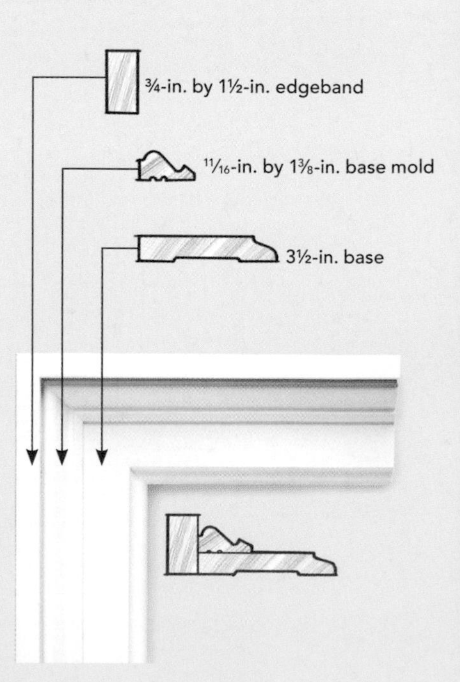

FACING PAGE: SIDES FIRST, THEN TOP. Measure between the sides to get the distance between the long points of the top piece. After fitting the top piece, glue the biscuits and miters, and fasten the top piece to the jamb edge.

RIGHT: START ADDING LAYERS. Measure along the outside edge of the flat casing to find the length of the outside edgeband at its short point. Press the edgeband against the wall, and nail it to the edge of the flat stock.

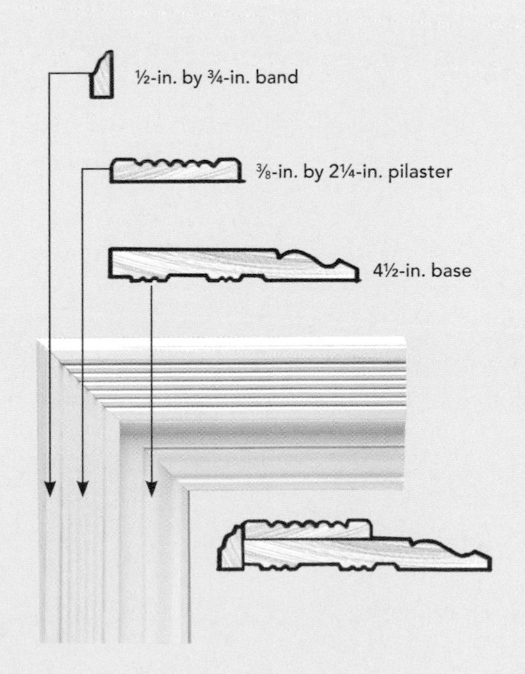

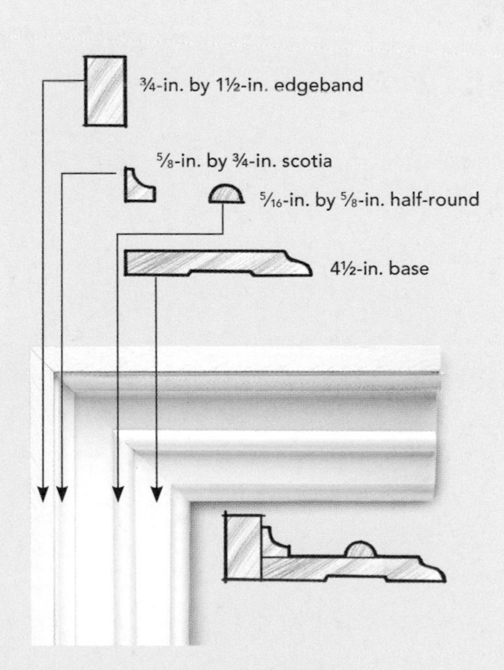

THE ART OF NAILING TRIM

Although the primary goal is to secure the pieces firmly in place, also think about hiding the nails whenever you can and spacing any visible nails as neatly as possible. It's also a good idea to locate the framing in the wall first, typically by probing through the drywall with a finish nail, but only in areas that will be covered by trim.

WHAT SIZE NAILS?

I use nail guns and typically shoot three different sizes of nails: 1¼-in. 18-ga. nails for small moldings, 2-in. 15-ga. nails for wood-to-wood nailing, or 2½-in. 15-ga. nails for nailing through drywall into framing. The standard nail equivalents are 4d, 6d, and 8d, respectively.

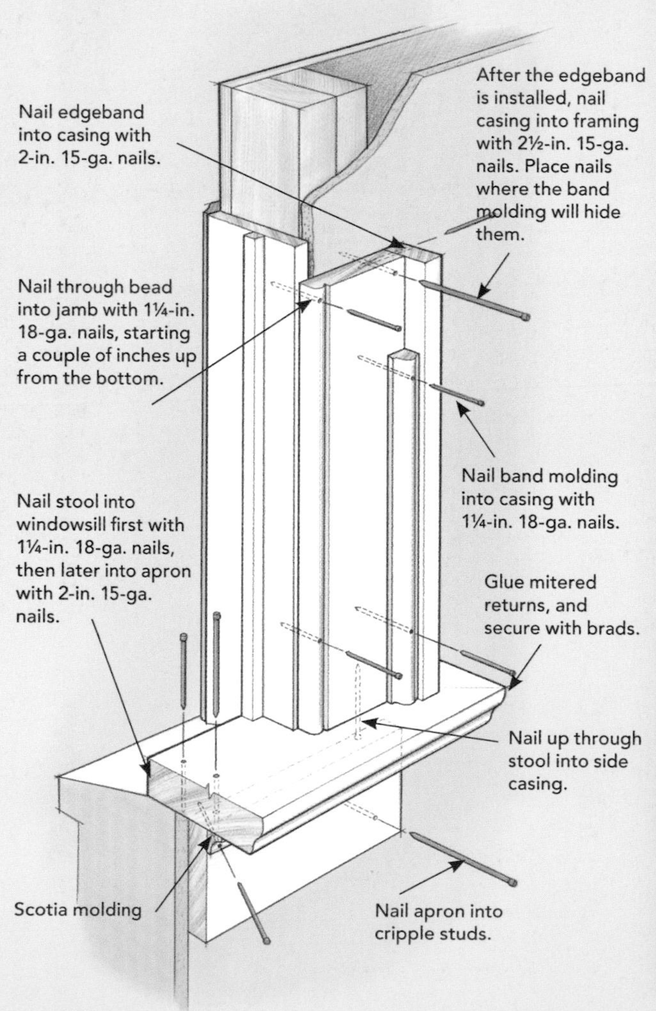

TRIMMING A MATCHING DOOR? IT'S EVEN EASIER.

THE PROCESS OF INSTALLING the built-up molding around doors is the same as it is for windows, only there is neither a stool nor an apron. If the finished flooring is in place, I simply measure from the flooring as if it were the stool. In this case, the finished floor had not been installed, so I set pieces of the flooring in place temporarily and used them to gauge the length and height of the door casing. After I've finished trimming the door, I pull out the flooring pieces.

THE APRON WRAPS UP THE JOB. With the decorative trim already attached, the apron lines up with the pencil marks made when the stool is laid out. Holding the apron tight to the underside of the stool, nail it to the framing.

Trimming a Basement Window

BY CHRIS WHALEN

Finish carpentry is the art of making rough stuff look good. Even trimming a window can be a challenge because it's usually complicated by poorly aligned framing or uneven drywall. If things go well, you can tenderize the drywall with a hammer or shim the window into alignment. If not, the window jambs might need to be planed, the casing tweaked, or the miters back-beveled at odd angles. In the end, a bead of caulk is often needed to disguise the solutions.

Multiunit windows in thick walls, such as the basement windows featured here, are prone to even more problems. For starters, even if the windows were installed plumb, level, and square, they might not be parallel with the finished wall surface, meaning that the side jambs need to be tapered. Second, the individual units might not be installed in a straight line, meaning that the stool needs to be tapered. Third, access between the window and the interior-wall framing could be limited, which reduces options for attaching extension jambs.

Identify the problems

The three window units here are in an 8-in.-thick concrete wall. A 2×4 wall covered with drywall sits inside. Before casing is applied to a window like this, the jambs and the stool need to be extended.

The first thing I do is determine how the window sits in relation to the drywall. With a multiunit window such as this one, I place a long straightedge along the top and bottom jambs to determine if the units are in the same plane and at the same elevation. In this case, the windows were at the same height, but the center unit was pushed out in relation to the flanking units. Next, I straddle the corners of each window unit with a short straightedge on the drywall and measure from the window jamb. This tells me how wide the extension jambs will be and if tapering is required. For reference, I write the measurement on the drywall along the edge of the opening where the trim will cover it later. If the variation is less than ⅛ in., there's no need to worry about tapering the extension jambs or stool. This discrepancy can be taken up by tipping the casing slightly. If the difference is greater than ⅛ in., the jambs need to be tapered.

Solutions start with the stool

Many windows have factory-applied 2-in. extension jambs that make the window suitable for a 2×6 wall. For basement walls, you need to extend the side and head jambs even more. I do this with a simple offset biscuit joint (more on that later). This offset joint

SAFETY NOTE: The author usually wears safety glasses when using a nail gun, but he forgot to wear them for this photo. Please don't make the same mistake.

looks good on the jambs, but it's impractical for a stool. That's why I carefully remove the factory-applied stool extension and replace it with a new full-depth stool.

The new stool needs to fit between the rough opening in the framed wall while extending past the side casings. The overall length of the stool is the sum of the distance between the side jambs, the width of the casings, the casing reveals (typically ¼ in.), and the amount of overhang beyond the casings. After cutting the stool to length, I miter the ends so that the profile returns to the wall. The extension is biscuited and glued to the back of the profiled stool. When this assembly is dry, I scrape excess glue, sand, fill gaps, and sand again, making it ready to install.

Set the new stool in the opening, and check its fit. The width will likely need adjustment. Because the three individual window units weren't perfectly in line on this project, I needed to taper the stool in addition to notching around the mullions. I use a square and a scribe to measure and mark the notches and the ends of the stool extending past the window. After removing excess material with a jigsaw, I slide the stool into position for final scribing and planing.

(Continued on p. 125)

REMOVE THE NARROW STOOL EXTENSION AND BUILD A DEEP ONE

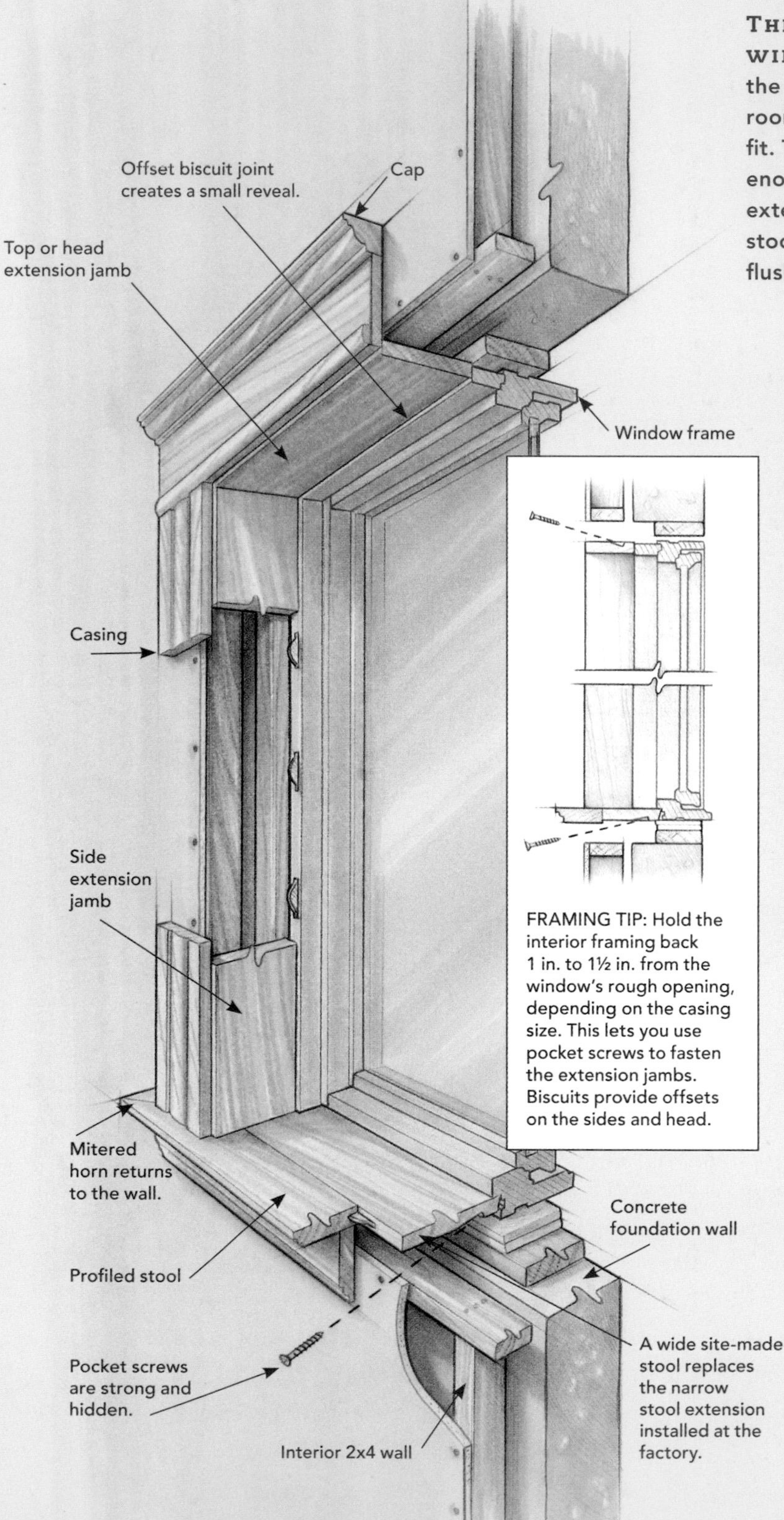

FRAMING TIP: Hold the interior framing back 1 in. to 1½ in. from the window's rough opening, depending on the casing size. This lets you use pocket screws to fasten the extension jambs. Biscuits provide offsets on the sides and head.

THE STOOL NEEDS TO BE WIDE ENOUGH to get past the drywall while leaving room to scribe the final fit. To get the stool deep enough, glue and biscuit an extension to the profiled stool, keeping the two parts flush on top.

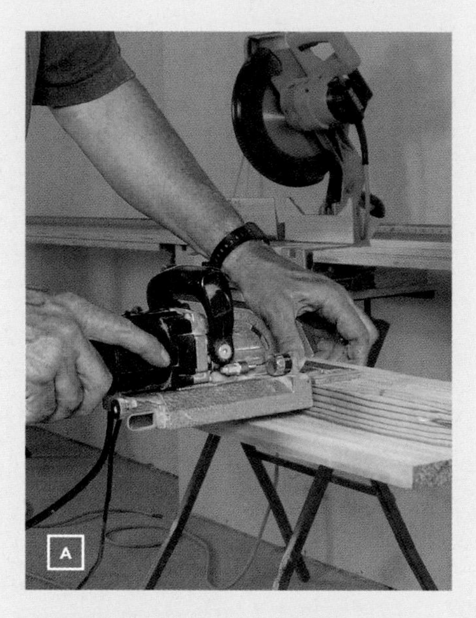

ASSEMBLE THE TWO-PIECE STOOL. I use biscuits and glue, then clamp the stool assembly overnight, making sure the tops of these two pieces are flush.

RETURN THE PROFILE TO THE WALL. I miter the returns at the end of the stool using two biscuits (stacked), glue, and blue painter's tape as a clamp.

A
B

FIT THE STOOL

WITH THE DEEP STOOL ASSEMBLED, scraped, puttied, and sanded, I turn to fitting. Ultimately, the stool needs to be tight to the window frame and drywall, and notched around the mullions. This begins with positioning the stool exactly parallel to the window and ends with a slight back bevel on the final cut. Rough- and final-scribing, cutting, and fine-tuning come between.

1. ROUGH-SCRIBE

MARK THE MULIONS AND HORNS. The depth of the notch and the amount I cut off the horns is the distance between the window frame and the stool (A). I square the notch lines at this depth and scribe the horns accordingly.

2. ROUGH-CUT

THE FIRST CUT IS THE DEEPEST. I use a jigsaw to cut the notches and horns, and a small circular saw to cut the length of the stool. The notches will be covered with trim later, so give yourself some wiggle room. The horns will be mostly covered but not where they return to the wall.

3. MAKE THE FINAL FIT

POSITION FOR FINAL SCRIBING. With the rough-cut stool back in place, I set my scribes to the widest gap. Next, I scribe the entire length of the stool, including the horns. This should make a perfect fit (C).

SOME CUTS MATTER MORE THAN OTHERS. The back edge of the stool is most important because it won't be covered by trim. To get a tight fit, I cut near the line with a saw, and then I ease the cut over to the line and back bevel with a block plane or sanding block (D).

4. INSTALL WITH POCKET SCREWS

BORE MANY POCKET HOLES. I put a screw every 6 in. to 9 in. on window stools for a strong connection because people often sit or lean on them.

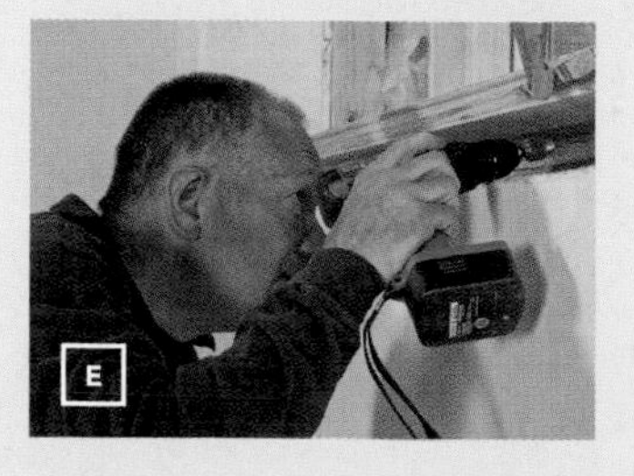

GIVE YOURSELF ROOM TO WORK. With space between the rough opening and the window frame, you can attach the stool extension with pocket screws (E).

FIT THE TOP AND SIDE JAMBS WITH AN OFFSET

For a great-looking joint that's fast to fit, I use a biscuit joiner with a clip-on offset plate. A ⅛-in. offset adds a shadowline to the profile and eliminates all the fussy fitting, sanding, and patching that a flush fit can require.

CLAMPS PROP THE TOP JAMB FOR SCRIBING. Just like the stool, the top extension jamb needs to be scribe-fit to all three window units. Don't get bogged down trying to get the exact length; it just needs to be long enough to land on the side jambs. What's important is that the top extension is parallel to the interior-wall surface when you scribe [1]. Scribe the back edge along the window frame [2], setting the scribes to the biggest distance that the front edge of the jamb sticks out past the drywall. Because the framing wasn't held back enough from the window, I had to face-nail the head and side extensions into the framing [3] rather than use pocket screws. The side jambs need to fit tightly top to bottom and also along their width [4]. If the framed wall isn't as plumb as the foundation wall (or as out of plumb), the board needs to be tapered. To get a tight fit top to bottom, I measure in two steps. First, I make a mark 20 in. up from the stool. Next, I measure down to the mark, and I add the two numbers together. This is more accurate (and faster) than bending my tape into a corner and guessing at the exact measurement.

The Lamello® Top 10 biscuit joiner has a clip-on offset plate.

1/8-in. offset
4

COMPLETE THE ASSEMBLY

THE TRIM DETAIL here was dictated by the trim in the existing house. I begin with the mullions, which need to fit tightly between the stool and the head extension. The side casings are cut ¼ in. long to establish the reveal for the head casing. The apron is installed last.

LONG HEAD CASINGS ARE A BIT TRICKY. I clamp the head casing in place and adjust the reveal to the head extension before nailing it off. I use a finish nailer with 2½-in. nails to attach the casing to the framing, and a brad nailer to fasten the casing to the extension jambs.

TIGHTEN THE STOOL. I use 2x blocks and shims to clean up the joints and make a solid stool. Last, I install the apron with mitered returns.

Finally, I bore for pocket screws, clamp the stool into position, and screw it to the window frame. I use a lot of screws (every 6 in. to 9 in.) because someone is going to sit on this window stool sometime in the future, and I don't want it to break.

Install jamb extensions

For the head and side jambs, I add a piece to the factory extensions using an offset reveal of about ⅛ in. The head jamb needs to be long enough to pass the side jambs, but it does not have to be fit to anything else. I cut it slightly longer than the overall length of the window. To scribe the head jamb, I set it in place with bar clamps and shims. Next, I measure at a few spots to determine what needs to be removed from the jamb stock, and I set the scribe and mark along the length of the jamb. I cut to the scribe line with a small circular saw, then use a power planer, a block plane, and a sanding block to adjust until the fit is acceptable. As with the stool, the process takes a couple of fittings.

At this point, I use a biscuit joiner to create a consistent offset or reveal between the extension jamb I'm making and the one applied by the factory.

On the project here, because there wasn't as much clearance between the window frame and the rough framing on top of the window as there was on the stool, pocket screws wouldn't work. Instead, after applying glue and inserting the biscuits, I shimmed and nailed the head-jamb extension in place, making sure it was square to the side jambs.

The only difference in installing the side jambs is that the length needs to fit precisely between the new stool and head jamb. Rather than bending my tape measure into a corner, I measure in two steps: up from the stool 20 in., then down from the head to the 20-in. mark. I then add the two numbers together. I cut the jambs to length and then to width according to the numbers written previously on the drywall. Finally, I fit the pieces and then biscuit, shim, and nail them in place, making sure they are square and tight to both the head jamb and the stool.

The rest is standard procedure

The last few steps of the process aren't much different than regular window trimming: Apply the mullion trim, casings, cap, and apron. I start with the mullions and work my way out. Using the same two-step measuring technique as I did with the side jambs, I measure the mullions, then cut and nail them in place.

I cut side casings to length, making them ¼ in. longer than the distance between the stool and head jamb, thereby creating a reveal at the head. After nailing them in place, I measure, cut, shim, and install the head casing and cap. Before installing the apron beneath the stool, I permanently shim and block the stool so that it is level, straight, and solid. I then make an apron with mitered returns on the ends the same length as the head casing, and I nail on the apron so that its ends are in line with the outside edges of the side casings.

Whether you're trimming a basement window or one in a double-stud, adobe, straw-bale, insulated-concrete-form, or any other thick-wall structure, these techniques ensure a quality installation for an appealing assembly.

Craftsman-Style Casing

BY TUCKER WINDOVER

Any cook will tell you that the fewer ingredients there are, the more important each one becomes. That's especially true with Craftsman-style trim treatments. This style has its roots in England's Arts and Crafts movement, which emphasized folk art and the workmanship of the individual craftsman. However, what developed in America as a residential Craftsman style was in many ways a reaction to overadorned Victorian homes built during the late 1800s and early 1900s. Gustav Stickley, who founded the periodical *The Craftsman* in 1901, became the purveyor of a style that strove to strip away excessive ornamentation and instead celebrate functionality and pleasing proportions. In the best examples of the period, the look was both simple and elegant.

That was then. Most of the time nowadays, my crew and I trim out windows and doors with off-the-shelf primed finger-jointed moldings. Sometimes, however, a client hires me to create a specific look. This requires me to shift gears and dig into my bag of tricks, which is the most enjoyable part of being a trim carpenter. Recently, I was asked to install Craftsman-style window and door casing.

Good examples spawn better designs

To help me settle on a design for this project, I did four different drawings of a window to try variations of Craftsman-style casing (see the drawings on pp. 128–129). This allowed me to experiment with the size and the proportions of the elements before I did any cutting.

Ancient Greek and Roman temples established a proportional system that relates the horizontal entablature to the height and diameter of the columns that support it. Similar proportions can be brought to bear on Craftsman-style trim for windows and doors (see the drawing on p. 129). While this is a good starting point, keep in mind that modern homes generally aren't designed with classical proportions.

Lacking a ready-made template, I looked not only at Stickley's work but also at casing treatments created by the Greene brothers, who were famous for the Craftsman-style homes they designed in California during the early 1900s. What I learned is that there's no such thing as a one-size-fits-all casing for the Craftsman style. A room's ceiling height and its window size affect decisions about casing dimensions. Formal areas such as a front entry door traditionally call for wider casing. In

closets and small rooms, bold trim can be overdone. The challenge is to let the casing add strength, mass, and mood without becoming overbearing.

Choosing and processing lumber is a big part of the job

Because this job involved stain-grade trim, I carefully selected the wood myself at the lumberyard. The vertical-grain Douglas fir I used on this job isn't a stock item at many lumberyards. The same can be said for straight-grained oak, another popular choice for Craftsman-style casing that will be stained or varnished rather than painted. When you are ordering and selecting boards, make sure that at least one edge is straight and square to register solidly against the tablesaw's rip fence. Most full-service lumber dealers will joint one or both edges of each board for a slight upcharge.

GUIDELINES FOR CRAFTSMAN DESIGN

WHILE THERE ARE SOME GOOD BOOKS on Craftsman style, no pattern book or carpenters' scripture provides exact measurements. Unless there's an architect involved, I design window trim myself based on my experience and a few guiding principles.

START SIMPLE, STAY SIMPLE

Use flat surfaces and square edges instead of molded profiles. Casing parts meet with basic butt joints. Subtle shadowlines are created by slight changes in thickness and overlapping elements.

WOOD GRAIN IS PART OF THE EFFECT

For this job, I chose vertical-grain Douglas fir. When I selected the lumber, I chose boards with tight growth rings and consistent color. Oak (either rift-sawn or quartersawn) is another popular choice.

BOLD, BALANCED PROPORTIONS ARE CRITICAL

To get it right, I often mock up a full casing treatment to make sure all the pieces work well together. Also, to enhance a custom look, I avoid off-the-shelf dimensions where possible.

THREE HEADER VARIATIONS

1. Head-piece frieze is thicker (1 in.) and overhangs the side and front by ¼ in. There is no fillet between the head and side casings.

2. Head and side casings have the same dimensions. A simple 1-in. by 1-in. backband is mitered at the top corners.

3. An angled molding is added under the cap. This detail also can be incorporated by beveling the edges of a thicker cap piece.

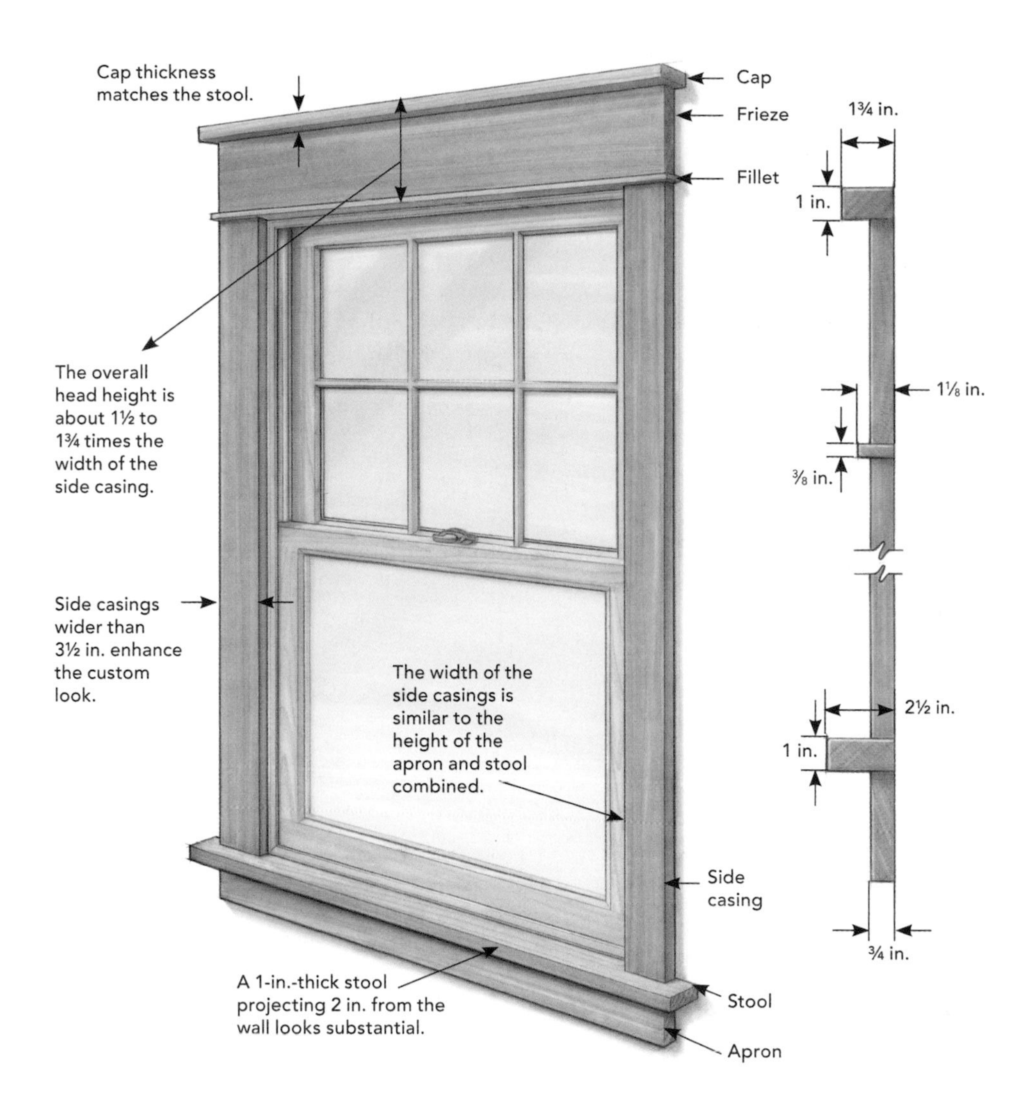

Once I'm on site, I separate wood into piles for stools, side casings, head casings, and other parts. Especially on a big job, it's easy to waste time counting, resorting, and restacking wood that isn't properly organized from the beginning.

After completing one window from start to finish and getting the client to approve the design, I can start mass-producing parts. It makes sense to keep the tablesaw's rip fence at the same setting until all parts that need to be a given width are cut. Only then do I move on to another rip setting.

I rip trim elements to within 3⁄16 in. of their final widths. Then I use a thickness planer to remove the saw marks on the edges. If the face has any chatter from the mill, I use the thickness planer to take 1⁄32 in. off the side that faces out. I identify the more attractive face of each piece and mark the back sides as I go. Once all the stock is separated into piles and cut to final dimensions, I do a whole-house takeoff and cut the pieces to length. As I work my way through the cutlist, I pay attention to consistency of color and grain pattern so that I can match pieces as needed.

The milling process leaves all the edges sharp. Because sharp corners are unpleasant to brush against and don't hold finish well, I ease the edges that will be exposed in the finished installation. A block plane

(Continued on p. 132)

SOLVING SOME PROBLEMS THAT LAND ON THE TRIM CARPENTER'S PLATE

Built-up compound, loose drywall, bulges in framing, and poor window installations are some reasons the window jamb and the drywall might not line up. Like it or not, these problems become the trim carpenter's headache. If the drywall surface and the jamb edges aren't made flush with each other, you'll be fighting the trim through the rest of the project, and it might never look right.

1. **APPLY FORCE IF NECESSARY.** Many times, the window is not set far enough inside the rough opening. Even if the trim and the siding have already been installed, you can often move the window by yanking it.

2. **DEMOLITION WORKS QUICKLY.** Before starting in with a hammer, hold the side casing in place and strike a line. Then move it an inch toward the window and score the paper with a utility knife. This provides a stopping point for the drywall paper when it peels away.

3. **PULL THE DRYWALL TIGHT.** A few screws can quickly solve the problem if the space between the drywall and the framing is a little slack.

THE STOOL COMES FIRST

The stool visually anchors the casings that rest on its top surface. It needs to fit well against the drywall, be square to the window, and be attached securely. Secure the stool temporarily to establish its length.

SCRIBE THE STOOL TO THE DRYWALL. Hold the stool in place, and with a pencil, scribe the profile of the wall onto the stool. The author uses a pencil for most scribing work, but you can also use a set of scribes or a compass.

HIDE THE FASTENERS IF POSSIBLE. If the window molding is substantial enough, screw the stool in place from the outside. Drive one 2-in. trim-head screw to secure the stool temporarily. Later, add one every 8 in. for a permanent connection.

CHECK FOR SQUARE. While the stool is temporarily secured to the window jamb, check for square. The cause of any tilt should be eliminated to avoid a visible gap between the stool and the side casing.

SIZE THE STOOL IN PLACE. On the first window, leave the stool horns long. With the stool temporarily attached, hold the side casing in place, and determine the stool length. A good rule of thumb is that the stool's side projection should match its front projection.

is a great tool for easing edges, but it's also possible to get the job done with 120-grit sandpaper.

With a simple design, the details matter more

For any custom job, I take a little more time prepping the opening and the materials. For Craftsman casings, this is especially true. You can't hide any flaws in a simplified design. If the window jamb and the drywall surface don't line up perfectly, the side casing cants out. This flaw can easily be seen where the casing meets the stool or the head. Also, if the window is even a fraction out of square at the corners, the butt joints where casing pieces meet show visible gaps.

Most window manufacturers have their own details for frame and sash, so the connections have to be worked out on a case-by-case basis. On this job, the Andersen® Windows moldings were thick enough behind the stool to allow trim-head screws to be driven from the back side of the window frame to secure the stool to the jamb (see the top right photo on p. 131). To lock this key piece in place, I nail it to the side casings (see the right photo on the facing page) and glue the top edge of the apron to the stool.

I use a rabbeted reveal block to mark the distance from the edge of the window jamb where the casing will land. This layout tool ensures a uniform reveal around the window. On this job, the jamb stock was thick enough to allow a ¼-in. reveal, which creates a pleasing shadowline. You can mark reveal offsets on the jambs with a utility knife or a sharp pencil. I often eliminate this step and simply measure to the top of the block (see the top left photo on the facing page).

I am careful about where I put fasteners. As a habit, I like to maintain consistent distances between nails (16 in. is typical) and to keep the nails in horizontally or vertically aligned pairs on the casing. One goes into the jamb; the other goes into the rough framing. Even if the nail holes are filled with expertly matched putty, a random nailing pattern is a visual distraction that's not acceptable on a stain-grade job.

A THICKNESS PLANER CLEANS UP FACES AND EDGES

I USE A THICKNESS PLANER to take a light finishing pass on the show face of every piece of trim. This removes any chatter marks or other surface irregularities while also ensuring uniform thickness. I also use my planer to clean up the edges of boards after ripping them slightly wider (about 3⁄16 in.) than their finished widths (see the photo below). Ganging boards together and running them through the planer gets the job done quickly and cleanly. After I run stock through the planer, I knock down the edges with a block plane and ease the edge of exposed end grain with 120-grit sandpaper.

CASINGS NEED A CONSISTENT REVEAL

Subtle details have a surprisingly dramatic impact on stain-grade trim. That's why it's important to establish a consistent reveal for the casing and to pay attention to where nails are driven.

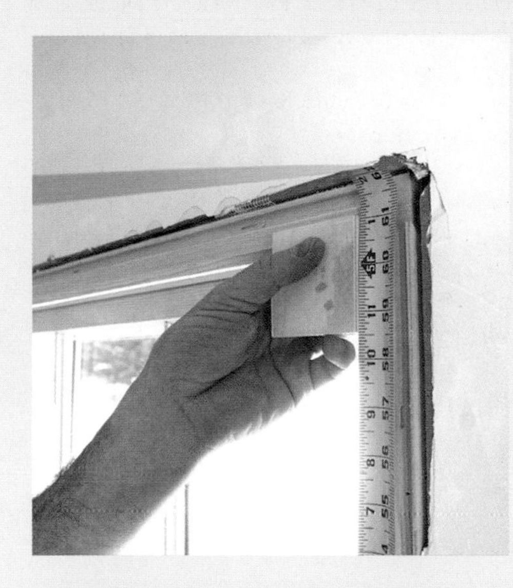

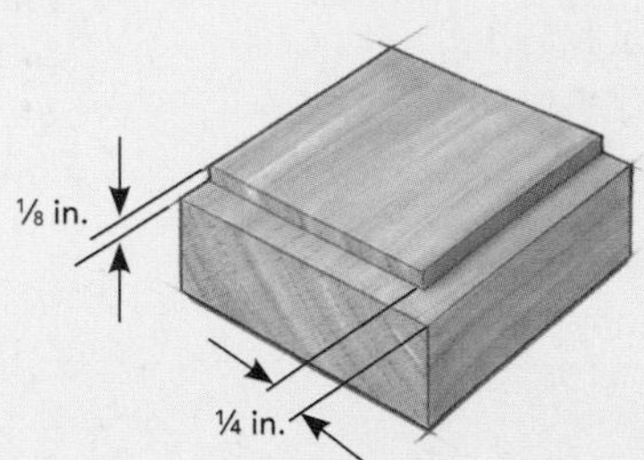

MEASURE TO THE TOP OF THE BLOCK. Rabbeted as shown in the drawing, a reveal block indicates where the top of the side casing needs to be.

MOVE OBSTRUCTIONS IF NECESSARY. Use a scrap of side casing to ensure that switch boxes and outlets aren't in the way. The box shown here is just far enough away that the author doesn't need to call the electrician.

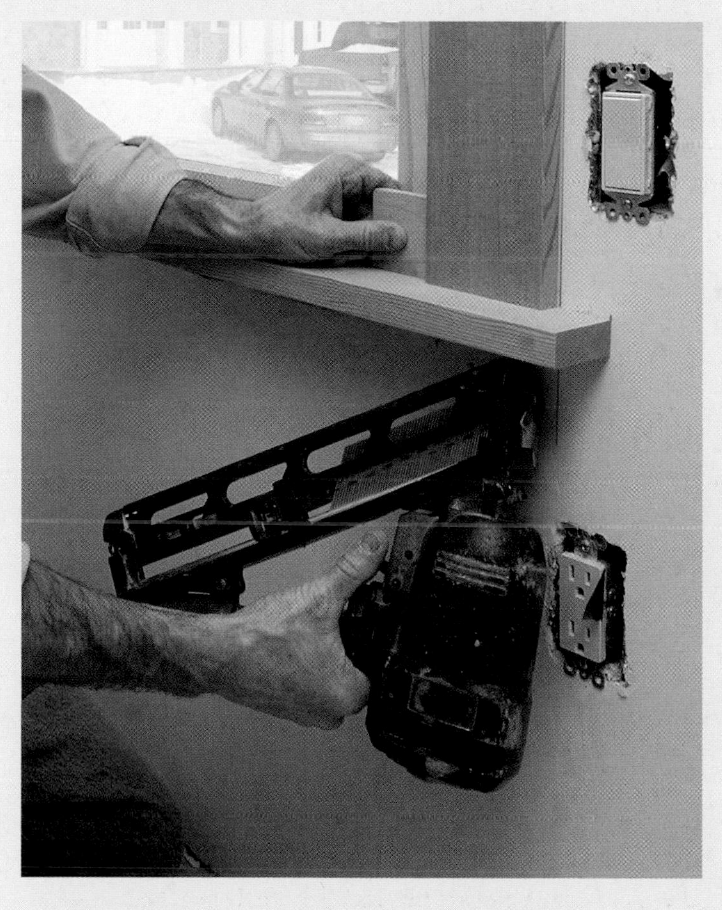

MAKE THESE NAILS DISAPPEAR. While holding the reveal block in place, secure the bottom of the casing from the underside of the stool. This eliminates two visible nail holes and helps to keep the joint tight.

I used 2-in.-long 15-ga. finish nails to install this trim. Thinner 18-ga. nails are too liable to bend if they encounter wood grain that leans toward the face of a board. You don't want to have to pull errant nails on a stain-grade project. I adjust my nail gun to set nails just over ⅛ in. below the surface. Any shallower, and the wood putty used to fill the nail holes might not stay in place.

Many times, especially during production-oriented work, the final walk-through can get the short shrift. On a custom job, however, I make a point of doing a thorough walk-through. One unfixed mistake can unnecessarily sour a client's perception of the whole job.

FRIEZE AND APRON ALIGN WITH SIDE CASINGS. Pencil lines on the wall extend the side-casing layout, making it easy to size and position the frieze board and apron correctly. The frieze is part of a three-piece header assembled on the bench.

ADD A THIN FILLET AND A FAT CAP. Once the frieze is cut to length, nail the fillet along its bottom edge and the cap along the top. The thickness of the cap matches that of the stool.

KEEP THE APRON CLEAN. Just as when you're installing the side casings and the header assembly, nail the apron carefully. The nails are best driven in pairs, aligned vertically on the apron and header, and horizontally on side casings.

Replace an Old Entry Door

BY EMANUEL SILVA

I've replaced dozens of rotten entry doors in my time as a carpenter. Unfortunately, most of those rotten doors never had a chance in the first place. In my opinion, the single largest cause of failing doors is improper installation and flashing.

Installing a door so that it's airtight and sheds water is imperative. Over the years, I've adopted a system that makes door installation easier, more accurate, and extremely weathertight. On this particular project, rot was less of an issue than aesthetics. The homeowners simply wanted a better-looking door. However, it's important to replace any material that shows even the slightest bit of rot before starting any of the sequences shown here. A solid substrate yields a flawless finished product.

Trim out the new door

Perhaps the greatest advantage to this approach is that you can trim the new door before you remove the old door. I find it easier and faster to attach trim—especially intricate dentil-molding details—when the door is flat on a worktable. I use PVC when trimming out exterior doors. PVC won't split, crack, or rot over time, and it never has to be painted. I've tried several brands of PVC trim and have found Kleer™ (www.kleerlumber.com) to be the most pleasing to work with.

TRIM THE NEW DOOR before you remove the old door.

1. ATTACH THE TRIM. The jamb is pocket-screwed to the PVC door casing, after which trim elements, such as fluted pilasters, can be attached.

2. BACK-FLASH THE CASING. Flexible flashing seals the joint between the jamb and the trim. A strip of flashing with the paper left on the exposed adhesive side should extend 4 in. beyond the trim. The flashing will be woven into the housewrap once the door is installed. A single strip of flashing can accomplish both tasks on narrower trim.

3. BUILD A SILL THAT SHEDS WATER. Secure a piece of factory-primed beveled siding to the sill with spray-foam adhesive and screws to direct water out of the assembly (see the sidebar on the facing page).

4. CREATE A SEAMLESS BASE. Held to a plumb line on the sheathing, the first piece of flashing extends across the bottom of the opening. A slit on each end allows the flashing to wrap over the sill and extend up the face of the wall.

5. ADD BOW TIES TO SEAL THE CORNERS. Use a bow-tie-shaped piece of flashing, made by doubling the flashing on itself and cutting it into a triangle, to cover the critical area where the outer edge of the bottom plate meets the sill.

6. WRAP THE SIDES. Cover each side of the rough opening with flexible flashing. Extend the flashing down over previously applied layers as far as possible. A single slit allows the flashing to be wrapped inside the opening and integrated into the beveled sill.

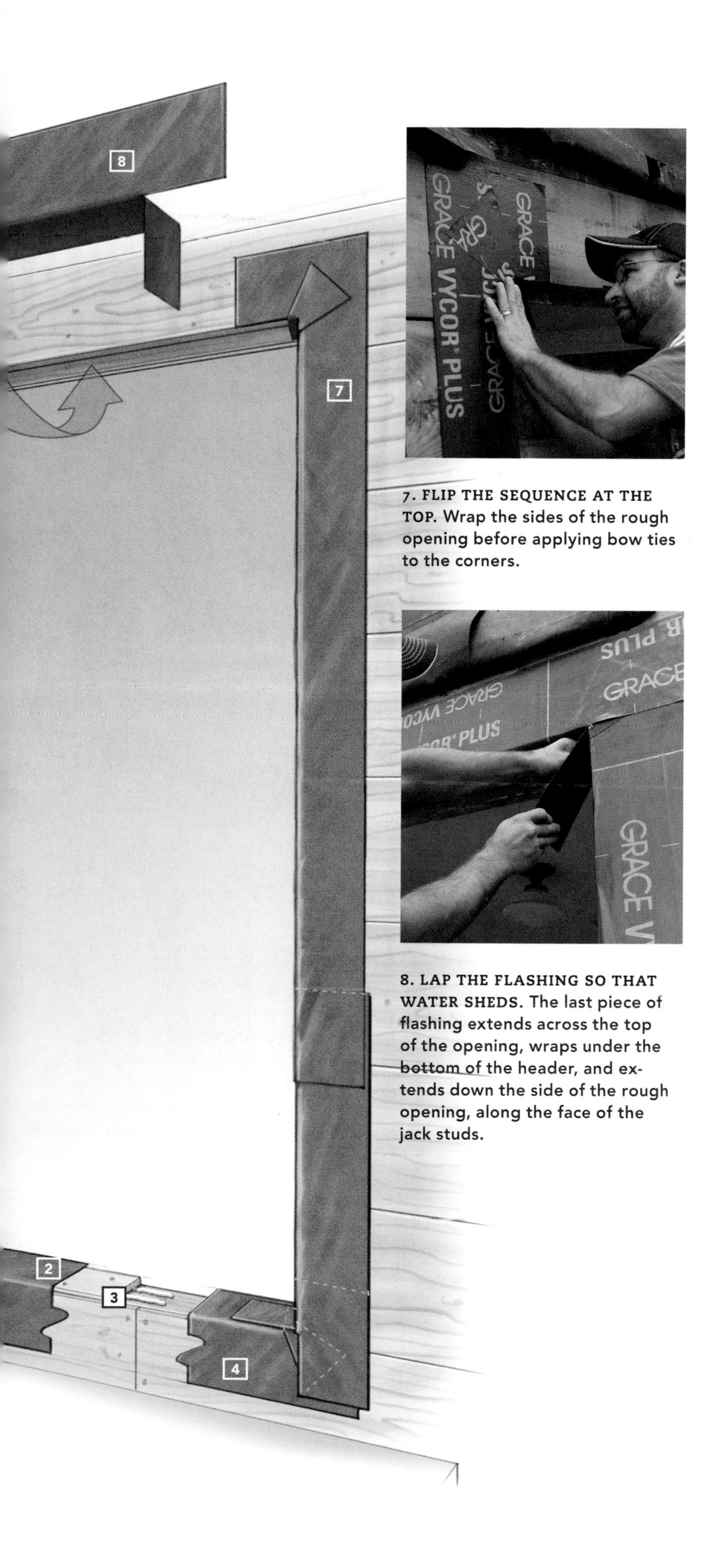

7. **FLIP THE SEQUENCE AT THE TOP.** Wrap the sides of the rough opening before applying bow ties to the corners.

8. **LAP THE FLASHING SO THAT WATER SHEDS.** The last piece of flashing extends across the top of the opening, wraps under the bottom of the header, and extends down the side of the rough opening, along the face of the jack studs.

SILL ALTERNATIVES

Beveled siding has long been the go-to material to create drainage at the door sill, but some argue that it doesn't provide enough support for the threshold and that the threshold can deflect and eventually separate from the jamb legs over time. In all my years as a carpenter, I've never had such a problem, and I believe the kick board offers enough support for the threshold at its outer edge. If you're unconvinced, though, or if your door manufacturer demands an alternative, here are two products worth looking into.

WEATHER OUT FLASHING®
A three-part sill pan has tapered ribs to offer full support for the threshold. The drain channels and back dam facilitate drainage.

Source:
www.weatheroutflashing.com

SURESILL™
A single, sloped sill pan (end caps optional) has front and rear support ribs. This sill pan also has a back dam to keep water out.

Source:
www.suresill.com

Flash the rough opening

Flashing the rough opening is the third line of defense against water infiltration. (Trim is the first, and back-flashing is the second.) Done correctly, the flashing protects the house's framing from water and rot. Failing to flash the rough opening correctly can trap water in the assembly and damage framing members and door parts.

Prep for installation

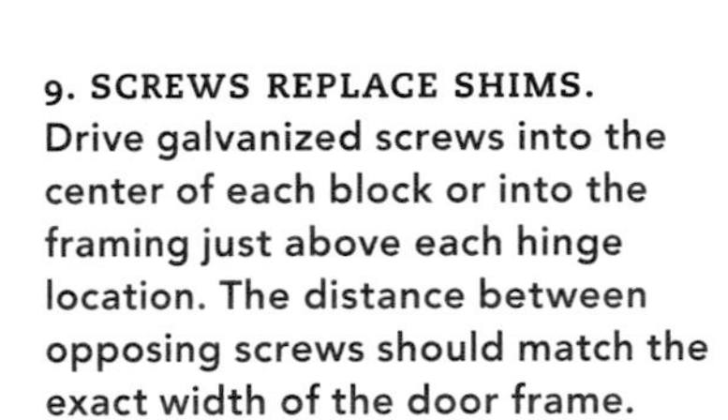

Typically, a door is put in place before wood shims are added between the door frame and the framing. Because the trim is already applied to the door frame on this project, I can't add shims from the front. Instead, I use screws as standoffs to plumb the jambs. There are a few benefits to this approach. First, plumbing screws before the door is installed ensures that the frame will be dead plumb. Second, it's faster to plumb screws than to fuss over wood shims. Finally, screws allow water to drain should it find

9. SCREWS REPLACE SHIMS. Drive galvanized screws into the center of each block or into the framing just above each hinge location. The distance between opposing screws should match the exact width of the door frame.

10. PLUMB UP. With the distance between the lower screws set, plumb up the side of the opening. Adjust the upper screws so that they are plumb with the bottom screws.

11. ADD A SITE-MADE GASKET. A rolled-up piece of flexible flashing keeps water and air from seeping between the bottom of the threshold and the sill. A few beads of caulk across the width of the sill aid in sealing and keep the threshold from squeaking.

its way into the rough opening. On this project, the rough opening was much wider than the door frame, so I padded with PVC blocking. If your opening is narrower, simply sink the screws into the jack studs.

Install the door

The scale of this project shows the applicability of this approach. No door is too big. However, no matter what size the door is, get assistance moving it into place. This will ensure that the unit seats into the sill gasket evenly. Also, a second set of hands will prove useful once you begin making minor adjustments to the fit of the door frame. Although it's not shown here, be sure to mount the door in its frame temporarily and to test the fit before you fully secure and flash the assembly. The final step in any door installation is to add low-expansion spray-foam insulation between the door frame and the rough opening. Your goal here is only to air-seal. Leave the bulk of the cavity empty to allow water to drain in the unlikely event that a flashing detail fails.

12. RESTORE THE DRAINAGE PLANE. Starting at the bottom of the door and working toward the top, peel away the paper backing on the back-flashing, and stick the housewrap to it, making sure that each piece is lapped over the piece below it. The back-flashing, which is sealed to the beveled PVC head casing and is woven into the housewrap, eliminates the need for metal cap-flashing.

13. FINAL FLASHING. Flexible flashing lapped up and around the door seals the joint between the housewrap and the back-flashing. It's completely overkill—and necessary. Apply housewrap tape to any seams or tears not covered by the flexible flashing before adding the kick board and bringing the siding tight to the trim.

Plumb Perfect Prehung Doors

BY GARY STRIEGLER

Several years ago, I walked onto a job site to find one proud employee. On his own initiative, he'd hung all the doors in the house alone—in less than three hours. Initially, I was impressed, but the 20 years that I'd been building houses tempered my excitement with skepticism. A little voice in my head said that I would regret not having given him a to-do list before I left—one so easy and so short that when I got back, I might find him in the pasture behind the house, practicing his Frisbee® throw with a dry cow patty.

The first door I checked was sufficiently nailed, opened freely, and didn't swing on its own when I let it free from my hands. But the rest of the doors had plenty of problems. Besides the fact that my proud employee didn't once use a level, he also failed to put shims in all the key places. What got me most, though, was that more than a few of the doors were swinging the wrong way. I'm all for getting things done fast, but accuracy is key when it comes to hanging doors. To minimize such mistakes in the future, I developed a door-hanging process that I could easily teach to my crew. It starts with making sure the right door ends up in the right opening.

Mark the rough openings

Ordering doors doesn't take much effort on my part because my salesman does it. But it does warrant a couple of hours of my time and attention to ensure that the doors show up without incident. That's why my salesman and I walk through the house room by room with the floor plans in hand before the electrician starts his rough-in. I like to get door orders out of the way before the drywall is installed to allow enough lead time for the order. At this stage, the walk-through is a good opportunity for me to catch any errors in rough-opening sizes or locations that my framers might have made. It also lets me visualize potential errors in door swing on the plans and to correct them as needed.

During this walk-through, I measure the rough openings to make sure they're 2 in. wider than the door size; this leaves ¾ in. for each jamb leg and ¼ in. of shim space on each side of the door. The door sizes usually already account for a ⅛-in. reveal around the door (for example, a 36-in. door will measure closer to 35¾ in.).

I write the size and swing of each door in permanent marker on the trimmer stud of its corresponding rough opening. This becomes the final size. I mark the plans if the size or swing has changed, and I make sure that my salesman makes the final list so that if a door shows up that doesn't match what's written on the trimmer, it's his problem to fix, not mine.

Marking the door swing on the trimmer stud also informs subcontractors who need to make decisions based on this information. Electricians need to know the door swing to locate light switches. HVAC contractors position return-air vents and feed registers according to door swing as well.

Finally, my hardwood-flooring contractor needs this information for certain rooms if he shows up before we hang the doors. Flooring transitions between wood and tile, for example, should happen under a closed door. If he knows the swing, then he can make the transition in the right place even if the door isn't installed yet. This process isn't ideal, though. I do my best to get the doors in before any flooring is in place. Then I can set the jamb legs directly on the subfloor and let the flooring contractor work around them as he goes.

Have a place to store the doors

There's nothing worse than not having a place to store all the interior doors for a house when they are delivered to a job site. That many doors—often 30 or so for the houses I build—take up a lot of space. So before the truck shows up, I make sure to have a safe, secure place to store them.

It is always a good idea to store millwork of any kind in a controlled climate. Doors can scratch walls, though, so if I'm storing them inside, I put them in rooms like the kitchen, where drywall damage will be covered by cabinets. No matter where the doors end up, they're stacked with the hinges facing away from the wall.

If I have to put the doors in the garage, I lay down heavy plastic or tar paper to keep moisture from wicking into the jamb legs and the door bottoms. On that note, though, I rarely store solid-wood doors in the garage because they're most prone to movement in humid conditions.

As they're coming off the truck, I inspect each door for damage. The damage I discover usually happens on the jambs. I look for splits in the jambs,

(Continued on p. 149)

HINGES FACE AWAY FROM THE WALL. On concrete floors, a vapor barrier protects doors from moisture. Here, the barrier is not in place yet.

SIZE AND SWING. Write the door size and swing on the hinge-side trimmer stud of each rough opening before the drywall is installed.

CLEAR ALL POTENTIAL OBSTRUCTIONS. Trim back the drywall on both edges of each trimmer stud using a rough-cutting handsaw.

SHIM THE HINGE SIDE PLUMB

PLUMB THE OPENING with one shim at each hinge location indicated by the black tape on a door-hanging level. Nail the bottom shim first, then the top, then the middle. Use long cedar shims, which are easier to handle and offer more adjustability because of their size.

1 Starting at the bottom-hinge location, nail the shim in place so that the thick end will face the hinge knuckle. Orient all three shims this way. As the drawing below shows, when the jamb is installed, the nail to the right holds the jamb tight to the thick end of the shim. The nail to the left will push the jamb slightly, as indicated by the blue arrow. This will splay the jamb a bit, minimizing the possibility that the door will bind. The movement is so slight that the eye will never pick it up.

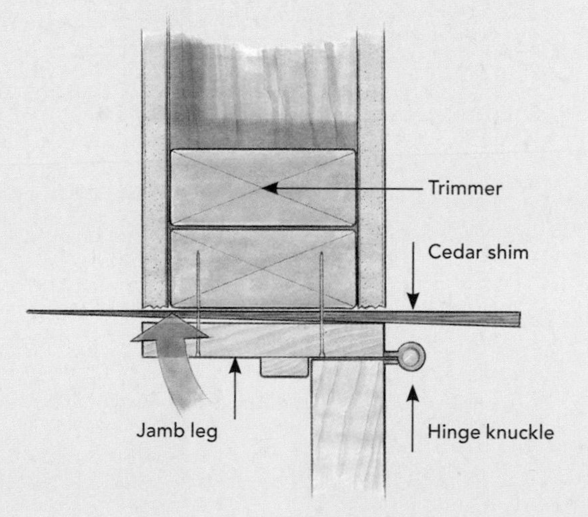

2 With the bottom shim nailed in place, slide the top shim between the level and the jamb until the level reads plumb.

3 Slide the third shim in until it just touches the level. If you push it in too far, you're likely to nudge the level slightly out of plumb. Trim each shim once it's nailed in place. If you're installing doors directly on the subfloor, place a shim on the floor at the hinge side to raise the jamb slightly.

IF THE FLOORING IS ALREADY INSTALLED, the jamb legs need to be cut. Using a level and a shim, determine the difference in height (if any) across the width of the opening. If the floor is level, trim both jamb legs so that the door will clear the finished floor by about ½ in. If the floor is out of level, trim that much more off the jamb leg on the high side.

1 OR 2 SHIMS,
2 NAILS
3
2 NAILS,
1 SCREW
1
1 OR 2 SHIMS,
2 NAILS
5
6 NAILS
6
2 NAILS
2
1 OR 2 SHIMS,
2 NAILS
4

SHIM AND NAIL, THEN REPEAT

Start by nailing the hinge side in place. Then shim and nail the latch side to create an even reveal (or space) between the door and the jamb, about the thickness of a nickel. Follow the sequence described below, adjust the door as needed, then finish the installation with one more nail through each shim. Don't shim or nail the head jamb; the casing will keep it in place.

1 Align the jamb so that it is centered between the drywall on both sides. Next, place one nail through the jamb and the shim just in front of the stop. Then replace the middle screw of the top hinge with one that's 2½ in. long. The top hinge bears much of the door's weight; this screw solidly anchors the door to the framing. Be aware that overtightening this screw can compress the shim and kick out the bottom of the door, causing it to catch on the latch-side jamb leg. If the door needs adjusting, this corner of the jamb is the last thing to be adjusted.

2 Center the hinge-side jamb at the bottom; then nail through the jamb and the shim just in front of the stop.

3 Center the top latch side, place a shim near the head jamb, and close the door. Adjust the shim to create a reveal about the thickness of a nickel along the side and top of the door. If the reveal is too tight, the door will stick in this corner. Then place one nail in the jamb just above or below the shim. Nailing above or below the shim locks it in place but allows you to adjust it until everything is working perfectly. Add a shim from the other side of the door if the gap is too big for just one shim.

4 Center the jamb leg, and add a shim behind the bottom of the jamb leg about 6 in. up from the floor. Check the reveal, and nail the jamb above or below the shim.

5 Shim behind the latch to even the reveal, then tack the jamb in place above or below the shim.

6 Add one nail, then close the door and check the reveal one last time. Adjust the shims as needed to tune the reveal. Then move to the other side of the door and close it to make sure it meets the stop. If adjustments are needed here, make them in this order: at the bottom of the latch side, at the bottom of the hinge side, at the top of the latch side, and finally at the top of the hinge side. Remember to remove the long screw if making any adjustments there.

MOVE THE DOORSTOP AS A LAST RESORT. If the door is slightly warped or if the jamb is slightly bowed, the only option might be to move the doorstop. Use a block of scrap 2× and knock the stop where a nail or staple attaches it to the jamb, not between them. Tapping the stop between nails or staples won't move the stop sufficiently and could split it.

edges that have been nicked, and evidence that the door frame came apart in shipping or from rough handling. Solid doors are hard to damage, but I have seen holes in hollow-core doors. All wood doors can be scratched pretty easily. Anything that isn't repairable (within reason) on site goes back with the truck, and my sales rep gets a phone call.

Assume the opening is not plumb

Before I hang even one door, I move all the doors to their respective rough openings. Once the doors are spread out, I start the hanging process by inspecting the rough opening. I make sure there aren't any obstructions like drywall, nails, or a long bottom plate.

The only time I use a level is to plumb the rough opening. I check the edge of the hinge-side trimmer stud first to make sure that the wall is plumb (i.e., the bottom plate is plumb to the top plate). If it isn't, the door will open or close on its own. If my level shows that the wall is within 3/16 in. over the height of the door, I leave it alone. If it's out of plumb, I tap the bottom plate as needed and toenail it to the subfloor to keep it in place.

Once the plates are plumb, I move to the inside of the opening. I use black tape to mark the hinge locations on my door-hanging level. If you don't have a long level, tape a short level to a long straightedge like a 3-in.-wide length of ¾-in. plywood.

I plumb the hinge-side trimmer stud with one shim at each hinge location. I tack the bottom shim in first, then move to the top, then the middle. Finish nails can work here, but sometimes the impact from the nail gun splits the shim or blows the nail right through it. Drywall nails are a good alternative.

Some people think it's necessary to double-shim here to counter the effect of the shim's taper, but I disagree. Unless the stud is twisted, I put the thick edge of the shim on the hinge-knuckle side. I use one shim to cock the jamb just enough to keep the hinges from binding. Besides, using one shim is quicker. Once the shims are in place, I use a utility knife to trim them flush with the drywall.

Tip the door into place

Although the hinge-side trimmer stud is ready, the opening is not done yet. Installing the doors before any flooring is in place is ideal because I don't have to cut any jamb legs. It saves me time at this stage, but the best part is that it ensures a tighter transition once the flooring is in place. But I don't count on a level subfloor.

Before I place the door in the opening, I put a shim on the floor with the thick part toward the hinge knuckle to raise the hinge side of the jamb slightly. Raising this jamb leg a bit ensures that I'll get the right reveal across the top of the door. Otherwise, if I set the door on the floor and this side is lower than the latch side, I have to cut the latch side once it's hung.

I tip the door into the opening and start by securing the hinge-side jamb leg to the trimmer stud. I often slip a wedge under the door to hold it open and in place while I'm working. I tack the top corner in place and replace the middle screw on the top hinge with a 2½-in.-long screw that fastens into the trimmer stud. Then I tack the bottom corner and move to the top of the latch-side jamb leg. I use one or two shims as needed in each location of the latch side, making sure that the jamb remains aligned with the wall plane.

Once the door is hung, I double-check the reveal around the door. If the reveal is even and if the door is working properly, I make sure it is closing fully against the doorstop. If it doesn't meet the stop, I adjust the jamb legs as needed. Occasionally, the stop might have to be moved with a wood block and a hammer.

When the door is operating to my liking, I finish by adding one or two more nails at each shim location along both jambs. Then I trim the latch-side shims. Although this is the end of the hanging process, it's also my least favorite thing to do. I've found that a sharp utility knife is ideal for the thin end of the shim and that a dovetail saw works best on the thick end.

CENTERED THE EASY WAY

If a door is centered between two walls, as at the end of a hallway, there should be an even space between the casing and the drywall on both sides. It's possible to hang the door as previously described by placing it in the opening and shimming it until it is perfectly centered, but that takes a lot of time. To save time and to ease installation, you can hang the door with the casing attached. This approach makes centering a tall door, like the one shown here, more convenient as well.

ATTACH THE CASING LEGS. After removing any nails that were holding the door in place for shipping, nail the casing to both jamb legs using 18-ga. brads. Make sure the casing is perfectly straight, and create an even reveal along the jamb. Leave the head casing off; it will be cut to fit once the door is hung.

CENTER THE TOPS FIRST. Tip the door in place; then center it by measuring the space between the drywall and the casing. Next, place one 15-ga. finish nail through the casing and into the trimmer stud to the left of the top hinge.

PLUMB THE HINGES. With a long level tight to the hinges, adjust the bottom of the jamb until it's plumb. Nail through the casing and into the trimmer stud. Adjust the latch side until the reveal is even, and nail it in place through the casing as well. Finally, install shims from the other side of the door, replace the top hinge screw, and nail the jamb as you go. Once the installation is complete, add the head casing.

Troubleshooting a Prehung Door Installation

BY TUCKER WINDOVER

The whole point of prehung doors is to save money on materials and labor, right? But even when the factory has mounted the hinges and assembled the jamb, a prehung door still can be tricky to install. After installing enough of these doors to spot the pitfalls, I have developed a series of techniques that explain away the mysteries and improve installations.

A prehung door assembly is really just a rectangle (the slab) swinging inside a rectangle (the jamb). I think of the door slab as a template for the jamb. The trick is to establish and maintain a consistent gap between the two, which should translate to a properly operating door.

Although prehung doors are far and away the most common setup found on a job site, there are some variations. The doors themselves may either be hollow core or solid core. Some pre-hungs (usually hollow core) are set in split jambs that fit together with a tongue and groove. These doors have casing already applied for a simple, one-person installation. The methods that I describe here, though, were developed to tackle the heavier solid-core doors.

Get started before the door delivery

The first thing my crew and I do before the doors arrive is trim back the drywall from the frame. To make distribution efficient, we mark the door size on the hinge side of the frame and indicate its direction of swing with an arrow. When the doors are delivered, we place each one as close as possible to its intended location.

The existing conditions of the floors and walls are usually more important than the level. For example, if I'm installing a closet door in the corner of a room and the adjacent wall is out of plumb, I typically follow the wall. A plumb door in this situation creates noticeably converging lines. At this stage of construction, I prefer to keep the lines running parallel. While one side of the door may be adjacent to an out-of-plumb wall, the other side's head jamb may need to be aligned with that of a nearby door. The solution is to split the difference between one side and the other. This situation happens a lot, and it's one of the reasons I prefer to hang doors with a partner.

The other thing that no longer surprises me is finding a wall—even in new construction—that's a bit wider (and occasionally narrower) than the jamb. By the time you add up the factors of a twisted

or misaligned stud, a proud nail head in the rough frame, or built-up joint compound on the drywall, the wall thickness can easily exceed the jamb width by $\frac{3}{16}$ in. or more. This can make the casing installation tricky, especially where the casing is mitered. Typically, when the jamb is more than $\frac{1}{8}$ in. shy of the drywall, I install a jamb extension to make up the difference because trying to make a neat miter joint over uneven surfaces is a time-consuming chore. The simpler solution is to center the head jamb between the sheets of drywall, with each side just a little shy of the plaster. Any straightedge that can span the corner of the rough opening will give a quick and accurate reading of the jamb's relation to the plane of the drywall. My colleague Alex and I take readings on each side of the wall, and we work together to split the difference.

It's a job for two

I achieve the most consistent results with prehung doors when I work in tandem with another carpenter. Although prehung doors are sold as a one-man job, there's too much to do on both sides of the door for one person to accomplish efficiently. Sometimes you're trying to install a door so that it's equal in height to an adjacent door, while at the same time centering the door in the opening, maintaining even gaps, shimming, and nailing. It's easier with two people, and once you've established a rhythm, it's quick. One person eyes the gap on one side, and the other person nails the jamb on the stop side. You don't need two carpenters to install every door, but when you've got momentum, it's nice to have those extra hands.

RETHINK THE PROCESS FOR BETTER RESULTS. **Instead of assigning one carpenter per door, use one carpenter on each side of the door. Communication and teamwork make installation faster than two people working solo.**

A BASIC INSTALLATION IN NINE STEPS

1. TWO CARPENTERS ARE BETTER THAN ONE
Alex has a 6-ft. level, and he stands on the interior of the door to establish a plumb hinge side and to check the gap around the door. The author stands at the door's exterior with shims and a nail gun. Alex can see when each jamb's gap is consistent and tells Tucker when to adjust or nail.

2,3. START AT THE HINGES
Shims are placed just under the top and bottom hinges. When the jamb is plumb, Tucker nails the jamb at the shims. Speed Squares® on both sides verify that the jamb is flush with the drywall.

4,5. NO FALSE READINGS
Remove the nail holding the door to the jamb and any cardboard shims stapled to the edge of the door. Now the door will operate as dictated by the position of the hinge jamb.

6. CHECK THE HEAD
If the gap is consistent across the head jamb, proceed to the strike side.

7,8,9. ESTABLISH THE GAP AT THE STRIKE SIDE
Shim the strike side at the top, behind the strike plate, and at the bottom. Check the gap, and nail off the jamb. Don't forget to check the swing, too. When the door is closed, it should hit the stop evenly down its length.

TROUBLESHOOTING A DOOR IN THE REAL WORLD

PROBLEM: A bowed jamb above the top hinge creates a gap that's too tight on the strike side.

SOLUTION: Drive shims at the hinge side of the head jamb until the gap is equal. Although this tight gap is more common on the hinge side, you also can use this solution to open the gap on the strike side.

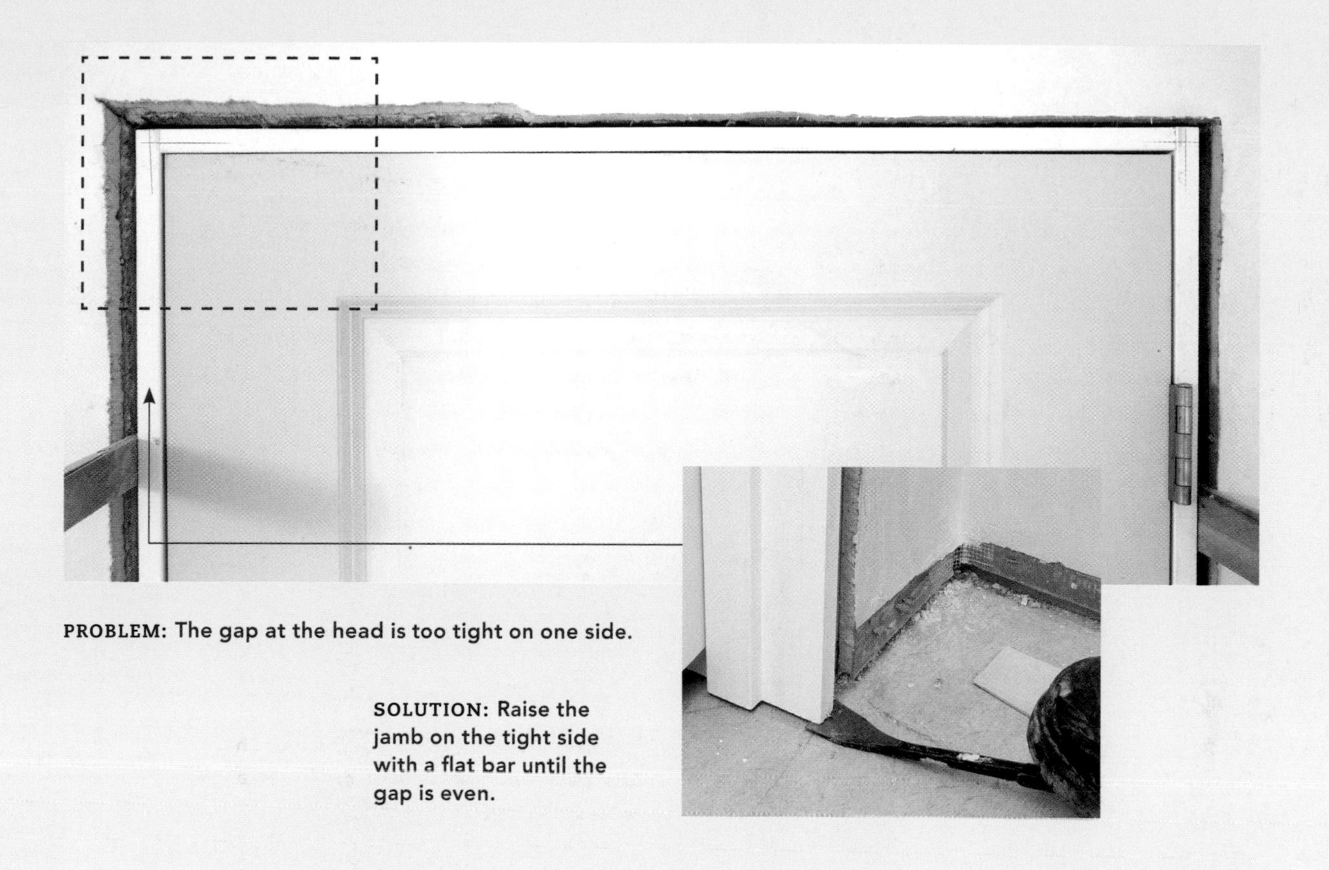

PROBLEM: The gap at the head is too tight on one side.

SOLUTION: Raise the jamb on the tight side with a flat bar until the gap is even.

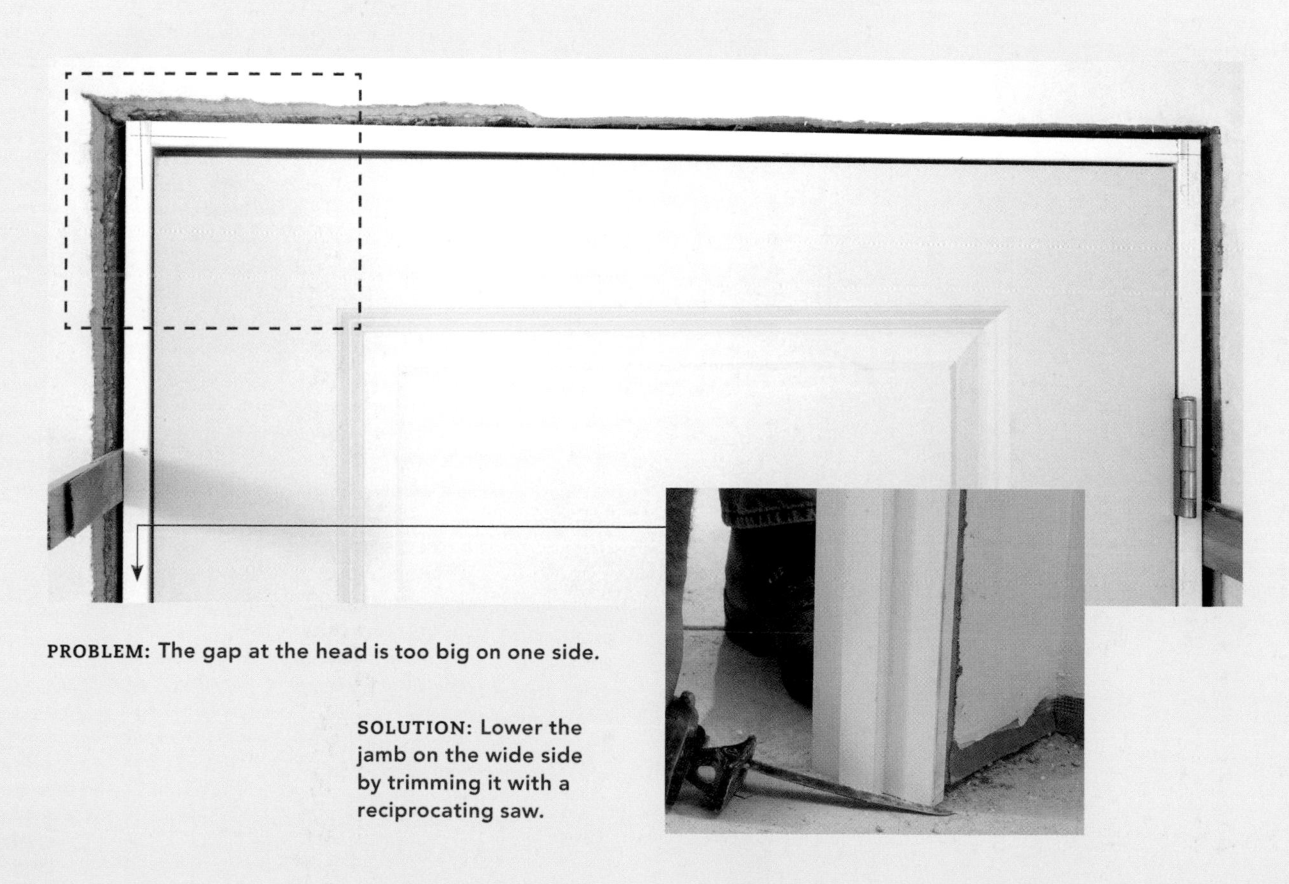

PROBLEM: The gap at the head is too big on one side.

SOLUTION: Lower the jamb on the wide side by trimming it with a reciprocating saw.

TROUBLESHOOTING A DOOR IN THE REAL WORLD

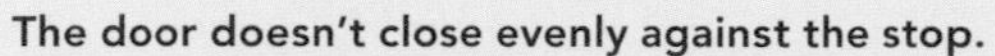
The door doesn't close evenly against the stop.

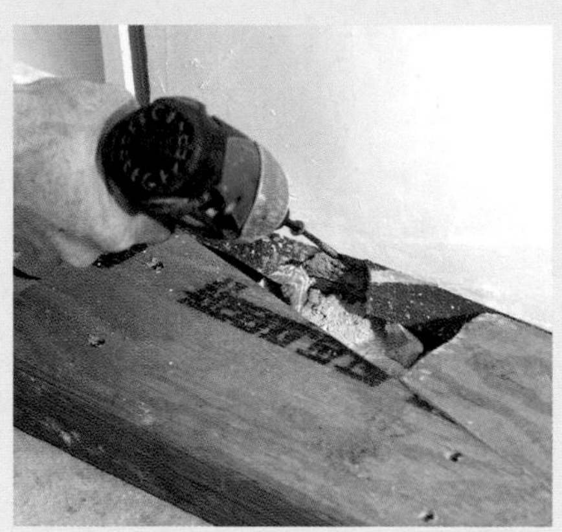

SOLUTION: Align the plates on each side of the rough opening. The door isn't closing evenly because the jamb legs are out of alignment. Unless the jamb legs can be adjusted back into alignment within ⅛ in. of the drywall plane, the solution is to move one plate. Cut two wedges from heavy stock, and screw one to the floor near the wall. Drive the second wedge into the gap until the door jambs are aligned (see the top photo). Cut the drywall back to gain access to the bottom plate, and drive a long screw through it into the subfloor to secure its new position (see the photo above).

PROBLEM: The door doesn't close properly because the hinges are binding or because the hinge-side gap is too big.

SOLUTION: Carefully bend the hinges. If the hinges are binding, the gap is too tight. To open the gap, use a crescent wrench to bend the jamb-side hinge knuckles toward the door handle. To close the gap, bend the knuckles away from the handle.

ALTERNATIVE SOLUTION: Another option for opening the gap is to insert an adjustable wrench or something of a similar size between the hinge leaves, then slowly close the door. Smaller items, such as a nail set, shouldn't be used because they can ding the door edge as the door is closed.

No-Trim Door Jambs

BY KEVIN LUDDY

Anyone who has ever done finish carpentry knows the frustration of trying to make perfect miters in the far-from-perfect world of new-home construction. The aggravation only increases when the molding is large and complex. When I was asked to do the trim work on a custom home recently, these concerns merged with the architects' desire for a feeling of simple elegance. The result was a door jamb that had an integral casing.

The cart comes before the horse

I made these jambs of finish-grade 2×6 whose inside face I double-rabbeted to create a minimal ¼-in. stop (see the drawing on the facing page). The backsides of the jambs have two dadoes that house the plasterboard seamlessly and eliminate the need for casing.

The nature of the plaster pockets and the fact that the jambs would be preassembled meant that they had to be installed before the wallboard was hung. In other words, I'd be installing finish work right after the rough framing and before the wallboard and finish-flooring materials went in.

To complicate matters, the builder didn't want the doors on site that early in the schedule for fear that they would be damaged. So I had to build and install

CLEAN AND SIMPLE. The minimal look of this finished door system hides the fact that the door jamb and casing are milled from the same piece of wood and installed before the wallboard.

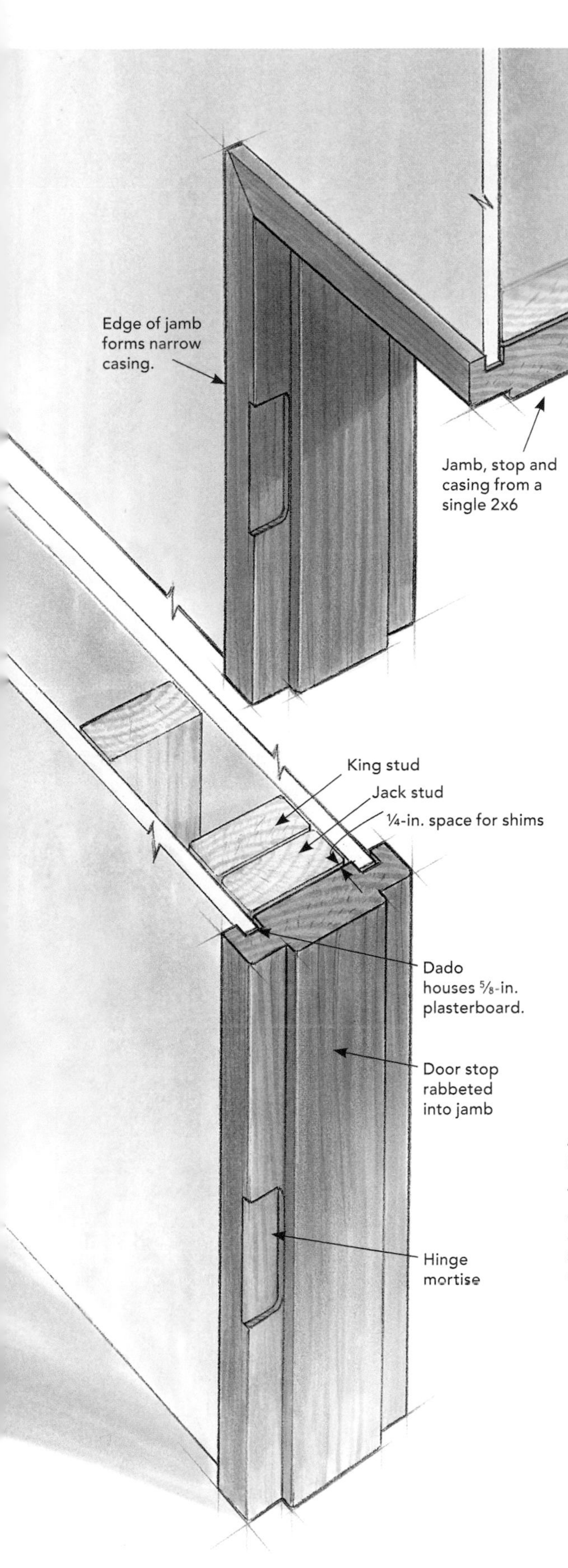

the door jambs using geometry only, with no doors to check the swing and stop. The whole concept was a little scary, but the image of a clean, simple, seamless jamb line gently rising off the wall plane was enough to fire my ambition (see the photo on the facing page).

Raw but refined materials

For a job of this size and scale (I had to make more than 20 door jambs in two different heights for doors of two different thicknesses), I'd normally get bids from mill houses to supply the jamb stock. However, in this case, dovetailing the schedules was of paramount importance. So I milled the door-jamb stock on site.

Besides the logistics, the biggest plus to milling the jambs on site was that I'd be able to handle all the stock up close and personal. I could select, match, and label each stick for its best location.

The first thing I needed was good stock: a large quantity of clear, vertical-grain Douglas-fir 2×6. It's generally better to be lucky than smart, and that rule came into full play when I called my supplier and found that he coincidentally had a truckload of 20-footers coming in shortly. I jumped at the chance and was able to make each complete jamb from a single board. I grain-matched the jamb pieces at each corner, giving each jamb a monolithic appearance as it wrapped its doorway.

A DOOR JAMB THAT DOESN'T NEED CASING

The trim for this door-jamb system is an integral part of the actual door frame. With the door jamb made of a single piece of wood, rabbets on one side form the stop, while plasterboard fits into the dadoes on the other side.

A SQUARING TABLE MAKES ASSEMBLY EASIER. **A large, flat table with blocks to keep the head jamb square to the sides helps to make assembly go quickly and smoothly.**

Milling the stock

After sorting and labeling the stock, I cut it to the rough head and leg lengths needed for each jamb. Milling the stock was basic. I received the stock S4S (surfaced four sides), and it needed no further dimensioning.

First, I cut double rabbets on the front sides to form the stops, making two passes through the tablesaw for each rabbet. I dadoed the plasterboard pockets on the tablesaw, using stacked dado blades set at 11/16 in. This slightly oversize dado comfortably accommodates the 5/8-in. plasterboard. I kept the slight factory roundover on the dado side of the boards, and I sanded the inside edges to match.

When the milling was finished, I sanded and sanded and sanded, slowly and deliberately. And when that was done, I sanded all the boards a little bit more. The biggest problem was the tendency of Douglas fir to rip out in long splinters at the worst possible places. All that I could do to remedy this situation was to monitor the feed rates carefully when I was dadoing and to assume that there would be some loss. Once the stock was milled, the builder had it all finished with water-based lacquer.

Assembling the jambs

Next came hinge-mortising and jamb assembly. The basic routine was the same for each doorway. First, I

BECAUSE THE JAMBS HAVE TO GO IN BEFORE THE FINISH FLOOR, a laser was used to set the finish-floor height. A small block attached to the jack stud supports the jamb at the proper height during installation.

PRE-SHIM FOR A PRECISE FIT. After the hinge locations are marked on the rough opening (see the photo above), shims are installed at each location so that the jamb can be installed perfectly plumb (see the photo at left).

mitered the jamb legs, leaving them a little long to allow remitering for grain-matching to the head jamb.

Then I mitered the head jambs, again adjusting for grain match. When I was satisfied with the grain match at both corners, I cut the legs to their finish lengths.

The next step was mortising for the hinges. I made a pair of shop-built hinge jigs (one for each door height: 6 ft. 6 in. and 6 ft. even) for this job. Making hinge-mortise jigs is a topic worthy of its own article, but I prefer my shopmade medium-density fiberboard jigs to commercially available jigs because I can customize them to the specifics of the job and because my jigs are more rigid.

After mortising for the hinges, I carefully assembled the jamb set for each door on top of a simple

WAIT. THE PLASTERBOARD ISN'T INSTALLED YET. Because of their design, these jambs require that the plasterboard be hung after the jambs are installed.

KEEP THE JAMB IN LINE. A stick the same width as the plasterboard is inserted into the rabbet to keep the jamb aligned with the framing while it's being attached.

PERFECT GEOMETRY. After the first jamb leg is plumbed and attached, a spreader stick ensures that the other leg is straight and plumb.

squaring table (see the photo on p. 160). I joined each corner with carpenter's glue and five or six 2½-in. #8 self-tapping wood screws.

Extra attention to framing simplifies installation

Arguably the most important step in the process was careful preparation of the rough door frame. I worked with the builder to bring the rough openings closer to the actual unit dimensions (only ¼ in. of shim space on each side) and to correct the king and jack studs for plumb, both of which were well worth the time spent.

READY FOR THE DOOR. When the plasterer was finished, the author installed the hinges on the door jamb and doors before the final fit.

I also used a laser level to establish the finish-floor heights in each room and marked the heights on the jack studs (see the top left photo on p. 161). I screwed plywood blocks to the rough framing for the jambs to sit on, which ensured a level installation if the jamb legs were of equal length.

When the opening was ready, I began jamb installation by marking and attaching shims perfectly plumb with each other at each hinge location (see the top and bottom right photos on p. 161). Next, I set the assembled frame (making sure it was labeled for that opening) onto the plywood blocks. I set the width of the jamb at the bottom using a spreader stick (see the photos on the facing page).

To set the jamb relative to the framed wall, I used a piece of wood the same thickness as the plasterboard as a gauge. I gave the jamb a final check of plumb, level, and square, cheating and tweaking as needed. I predrilled and then drove 3-in. #10 screws at each hinge location to attach the first jamb leg. Then I shimmed and attached the second leg.

I drove fasteners through the hinge mortises whenever possible to minimize the number of visible fasteners. On double doors, I was able to screw both legs to the jamb. On single doors, I screwed through the latch plate and drove a couple of 10d finish nails to finish attaching the jamb.

Doors: The final frontier

Protecting the finished jambs through board hanging and plastering was no small feat. The builder came up with an ingenious solution: vinyl J-channel that fit snugly over the jambs' exposed-trim area. The rest was taped heavily.

The plastering went fairly smoothly, although I regret not giving the board hangers a quick overall lesson before they started. They broke one jamb and twisted a couple more; however, a polite-but-firm pow-wow eliminated any further damage.

The doors' installation was the real time of reckoning. With the jambs plastered in, little or no adjustment was possible. The first step was to measure the door and the jamb opening carefully for any irregularities.

The doors were the heavy, solid-core flush variety, so minimal handling was a goal. I considered two trips to the bench to be perfect. On the first trip, I cut and beveled the door to 1⁄16 in. less than the finished opening rather than the standard 3⁄16 in. to make sure I had plenty of door left for fitting.

Next, I mortised for hinges, installed the door, and tested the fit. In most cases, substantial planing was needed, so I scribed the door, returned it to the bench, planed it to my marks, then reinstalled it. On a couple of doors, I did minor work on the jambs to make the door hit the stop evenly. A couple of doors needed only minor planing after the initial fit, which I was able to do in place, making them one-trip doors. My back loved me for that.

Case a Door with Mitered Trim

BY TOM O'BRIEN

There are almost as many ways to case a doorway (or a window frame) as there are carpenters. In all cases, the keys to success are making sure that the corners of the jamb are perfectly square, that your miter saw is cutting accurately, and that you assemble the miters carefully (get them right and tight) before nailing the rest of the casing.

To permit the casing to lie flat, the jambs should be flush with, or slightly proud of, the wall surface. Plane the jambs if they're too far out; extend them with thin strips of wood if they're too far in. If the drywall is proud of the jamb by ⅛ in. or less, knock it back with a hammer.

If you have only one or two doors to case, a 16-oz. hammer and a nail set will get the job done just as quickly as an air nailer.

PREP THE JAMBS AND CUT THE MITERS

LAY OUT FOR THE REVEAL. Use a combination square and a sharp pencil to scribe the reveal, the distance (typically $^3/_{16}$ in.) between the edge of the jamb and the casing. There's no need for a continuous line; simply scribe a dash every foot or two, making sure that the dashes meet in the corners.

START SQUARE TO STAY SQUARE. Check the jambs for square, and shim them if they're not. Otherwise, you may have to custom-fit the miter. Also, check the miter saw to verify that the blade is square to the table and to the fence; if so, the miter settings should be dead on. Refer to the owner's manual if adjustments are necessary.

CUT MITERS IN ADVANCE. Miter the side casings and one end of the head casing. Leave enough extra to cut these pieces to length later.

MARK IN PLACE, THEN INSTALL WITH GLUE AND FINISHING NAILS

INSTALL THE HEAD CASING FIRST. Align the mitered end of the head casing with the corner of the reveal, and mark the point where the far end meets the reveal. After cutting the miter, position the casing carefully, and tack it to the jamb with 4d nails; leave nail heads about ¼ in. proud in case adjustments are necessary.

MARK THE SIDES UPSIDE DOWN. Flip the side casing so that the mitered end touches the floor, and mark the point where the bottom intersects the top of the head casing. Make a square cut, and you're ready to install the side casing. A word of warning: If the finished floor is severely uneven, you may need to use scribes or a contour gauge to transfer the floor's profile to the casing before making the cut.

GLUE AND NAIL THE CORNER. Apply a generous amount of carpenter's glue to the end grain. Then carefully align the miter joint, and tack the side casing to the jamb with two or three 4d nails. Place the first nail about 1 in. from the corner. Blunt the point of this nail with a hammer to minimize the risk of splitting the wood. Secure the outside corner of the miter by driving a 4d nail up through the edge of the side casing into the head casing; drill a pilot hole for the nail to prevent knocking the whole thing out of whack.

EYEBALL THE SPACING. After the miters are fixed, the casing is nailed home. Use 4d nails to fasten the casing to the jambs; 6d or 8d nails are needed to secure the casing to the framing. Nails should be placed an inch or two from each end and the same distance from the hinges. Otherwise, space the nails 8 in. to 16 in. apart. Leave all the heads slightly proud of the surface; then use a nail set to drive them about ⅛ in. below the surface.

A JIG FOR MEASURING AND MARKING CASING

THE TRIM-LOC™ CASEMENT-INSTALLATION TOOL (Bench Dog Tools; www.benchdog.com) is essentially a triangle square that has been designed for casing. This compact tool fits easily in a nail pouch. Like a Speed Square, the tool has legs for marking 45° and 90° angles, but it includes extra features such as a gauge for scribing a 3⁄16-in. reveal. It also can be fastened to a workbench for use as a jig to transfer measurements accurately from the inside of a miter.

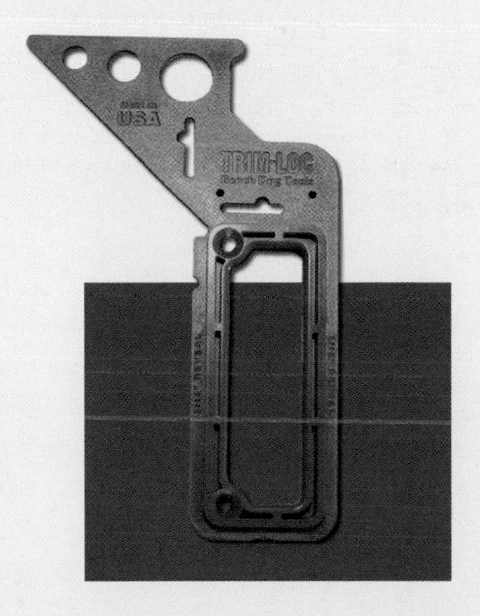

Turning Corners with Beaded Casing

BY SCOTT MCBRIDE

I suspect that beaded trim is common in older houses because it was attractive and easy to make by hand. A beading plane, unlike a wide molding plane, doesn't require much effort to push. It's an enjoyable task, if you're not in a hurry. A well-sharpened cutter makes a pleasant "snoosh" sound as it skates along, sending up slender straws of wood in its wake.

Beaded casings in old houses typically are joined with a combination butt-miter joint. The flat portions of the legs and head are butted at right angles, but the beads join in a miter (see the photo on the facing page). Flat casing with an ogee profile along its inside edge also can be joined this way. I don't know of an official name for this joint; I call it a butt-miter joint.

Hybrid joint combats seasonal shift

Why would you spend the extra time to make a fancy miter joint? Doesn't a plain miter work just as well? No, and here's why: Wood shrinks and swells almost entirely across the grain. The wider the casing in a plain-mitered joint, the wider the gap (see the drawing on the facing page).

In contrast, a butt-mitered joint—no matter the season or the width of the casing—shows only a small gap in the joint's mitered portion. The butted portion of the joint, where the legs meet the head casing, shows little or no gap because the length of the leg doesn't shrink. In addition, a biscuit inserted here and a finish nail driven through the face of the head casing into the framing further reinforce the joint, directing the cross-grain shrinkage of the head casing to occur from the top down. (Predrilling the head casing before nailing allows the wood to pull a little against the nail.)

Cut the legs first

As with any casing installation, I scribe a line on the edges of the jambs to gauge the margin, or setback, from the edge of the jamb. Next, I rough-cut the legs and stand them in position, scribing the bottoms of the legs to fit the floor or sill, if necessary. I mark the short points of the miters on the beaded edges of the casing legs where they intersect the scribed line on the head jamb. Next, I make a 45° miter through that point all the way across the leg (see the drawing on p. 170). Then, I square-cut the leg through the long point of the quirk, the groove that separates the bead from the flat part of the casing (see the drawing on p. 171). I mark a centerline on the face of each leg, then cut a biscuit slot into the top before

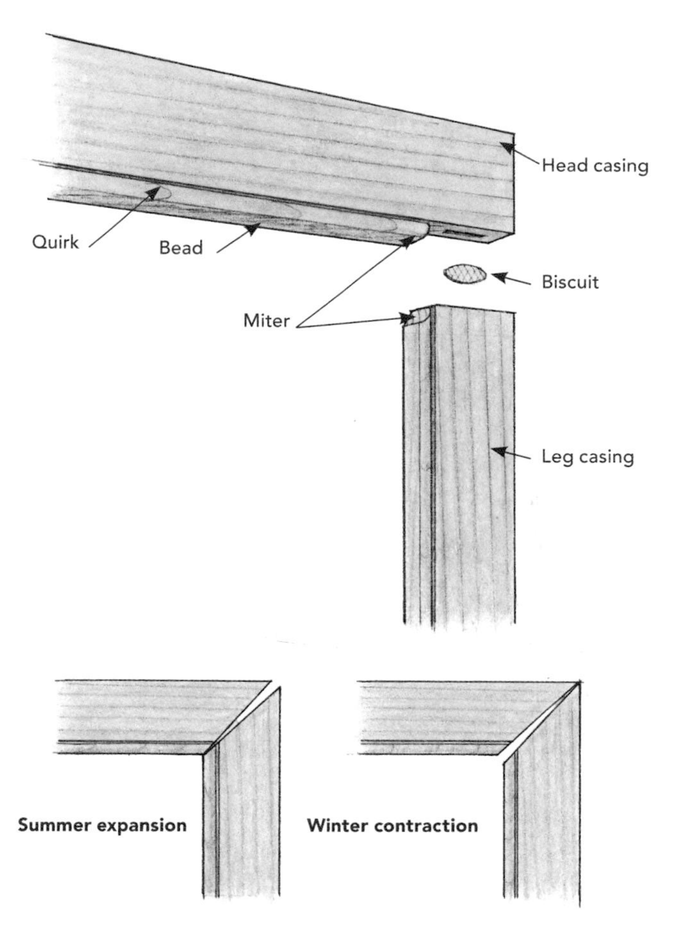

nailing them into place. For now, I leave out the nails in the top 3 ft. of each piece so that I can tweak the tops of the legs slightly when I fit the head casing.

Marking and cutting the head casing

Two cuts on the head casing are necessary to complete the butt-miter joint. One cut is a miter; the other is a rip along the quirk, which removes a short section of bead from both ends. To mark the head casing to be cut, it's best to place a longer than necessary piece of stock on top of the legs and mark it (see the photo on p. 172).

Mitering the head casing is the trickiest part of the job; if you're good with a handsaw you can just wing it, or make a simple jig to guide the cut. But if you have many of these joints to do—or if you're doing stain-grade work—it pays to tool up. A radial-arm saw works well on the miter cut, but because they weigh a lot, radial saws aren't exactly job-site friendly. A tablesaw is okay for short head casings, but the ones found over wide openings and mullioned windows are too long to wrestle across a tablesaw. So I came up with a jig that's worked out really well.

The butt-mitering jig is made from an old circular saw that's bolted to a carriage, which slides in a track. As the carriage is pulled, the sawblade passes through the casing's bead, which sits below in a trough similar to an old-fashioned miter box (see the drawing on p. 174). The depth of cut on the butt-mitering jig is set to cut right down through the quirk to the shoulder of the flat portion of the casing.

After making both miter cuts, I rip off the bead with a tablesaw, starting from the end of the stock and cutting until I get to the miter cut on each end. This final cut removes the short section of the bead and quirk, creating a square shoulder on the flat portion of the stock.

So that I don't have to readjust the fence constantly, I set up two tablesaws side by side for cutting opposite ends of the stock. This saves a lot of time if I'm trimming more than a few doors or windows (see the top drawings on p. 173).

With the beads removed, I set the head in place to see how it fits. If necessary, I pare the joint with a sharp chisel. When the fit looks good, I transfer the biscuit mark from the leg to the head and mark

MARKING AND CUTTING THE LEGS

MARKING RATHER THAN MEASURING is a fast and accurate way to determine where to make the cut for the butt-miter joint. Begin by holding the jamb leg in place; it should be a little longer than needed and can be scribed first to fit an uneven floor if necessary.

MARK THE BEADED CASING IN PLACE

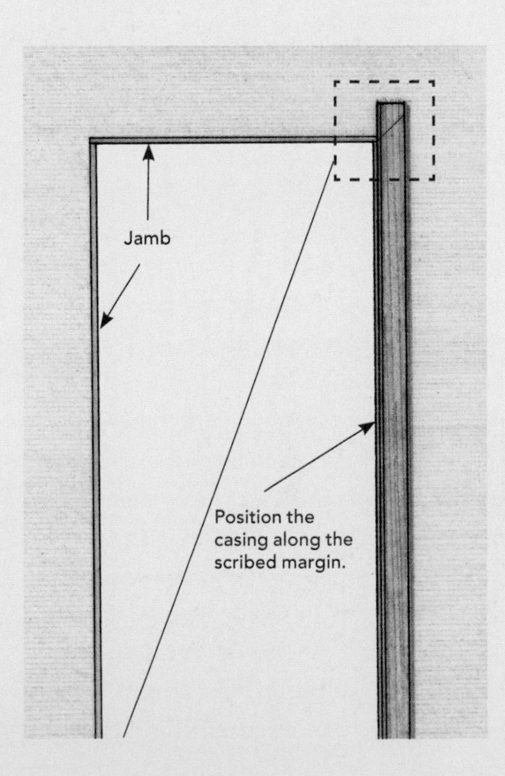

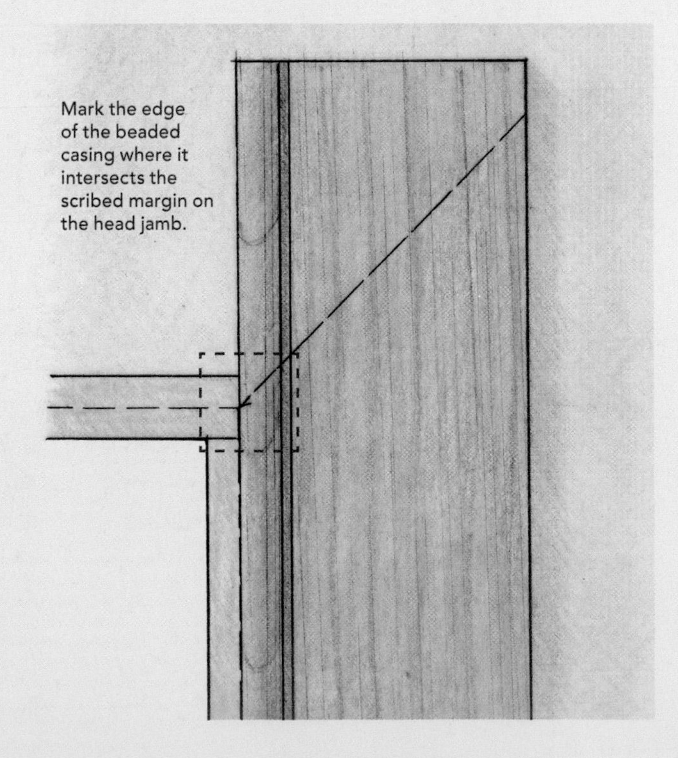

the ends of the head for trimming (see the top photo on p. 175). I can't use a biscuit joiner for slotting the head because the bead gets in the way. Instead, I use a self-piloting wing cutter chucked in a router. Then I glue a biscuit into the slot on each leg and nail the head casing into place. If necessary, I shim the back of the head casing so that the face of the head and leg casing lie within the same plane (see the bottom photo on p. 175). Finally, I run the backband around the outside edge of the beaded casing.

THE LEGS REQUIRE TWO CUTS

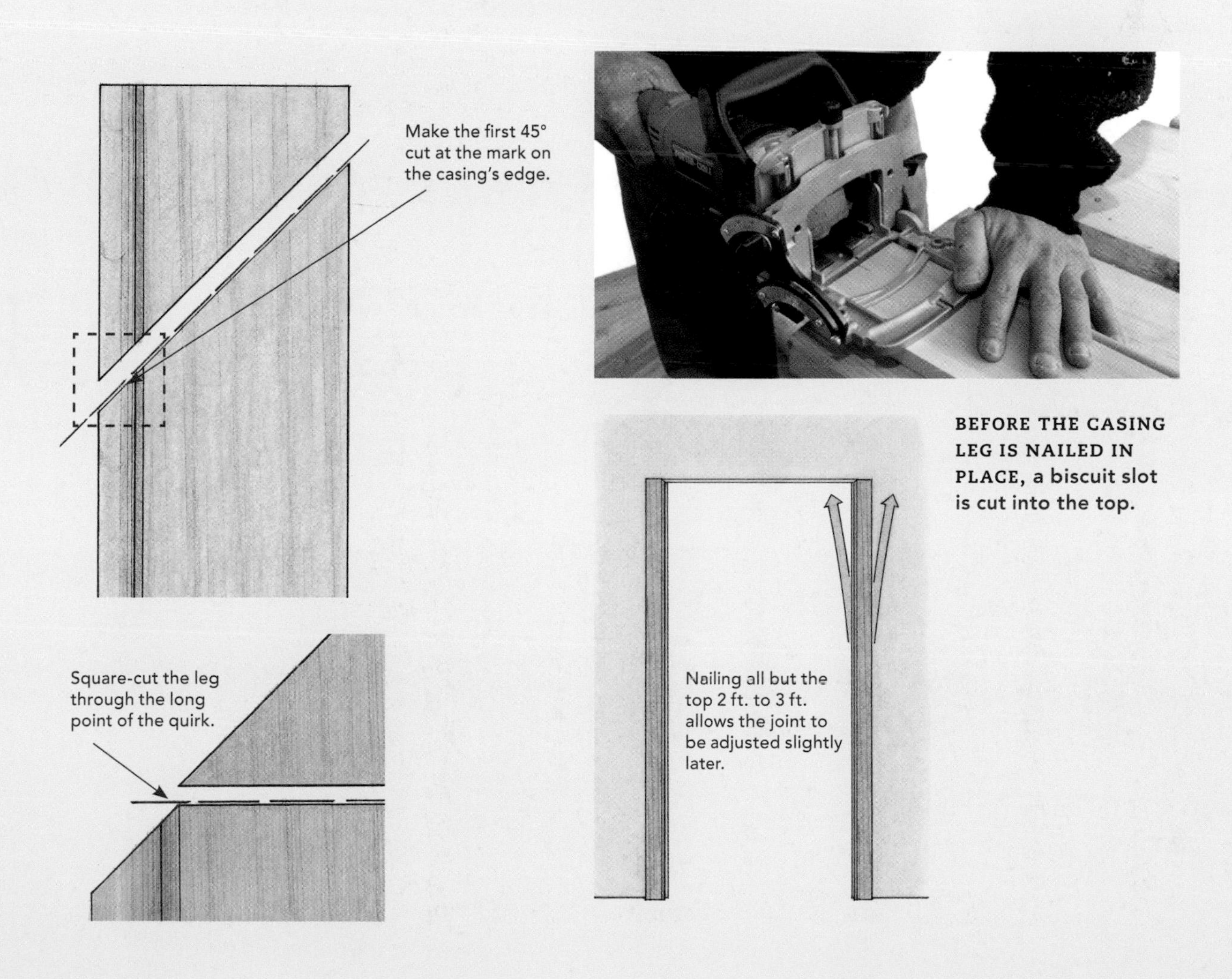

BEFORE THE CASING LEG IS NAILED IN PLACE, a biscuit slot is cut into the top.

MARKING AND CUTTING THE HEAD CASING

CUT THE HEAD CASING LONG, place it atop the legs, and mark it in place.

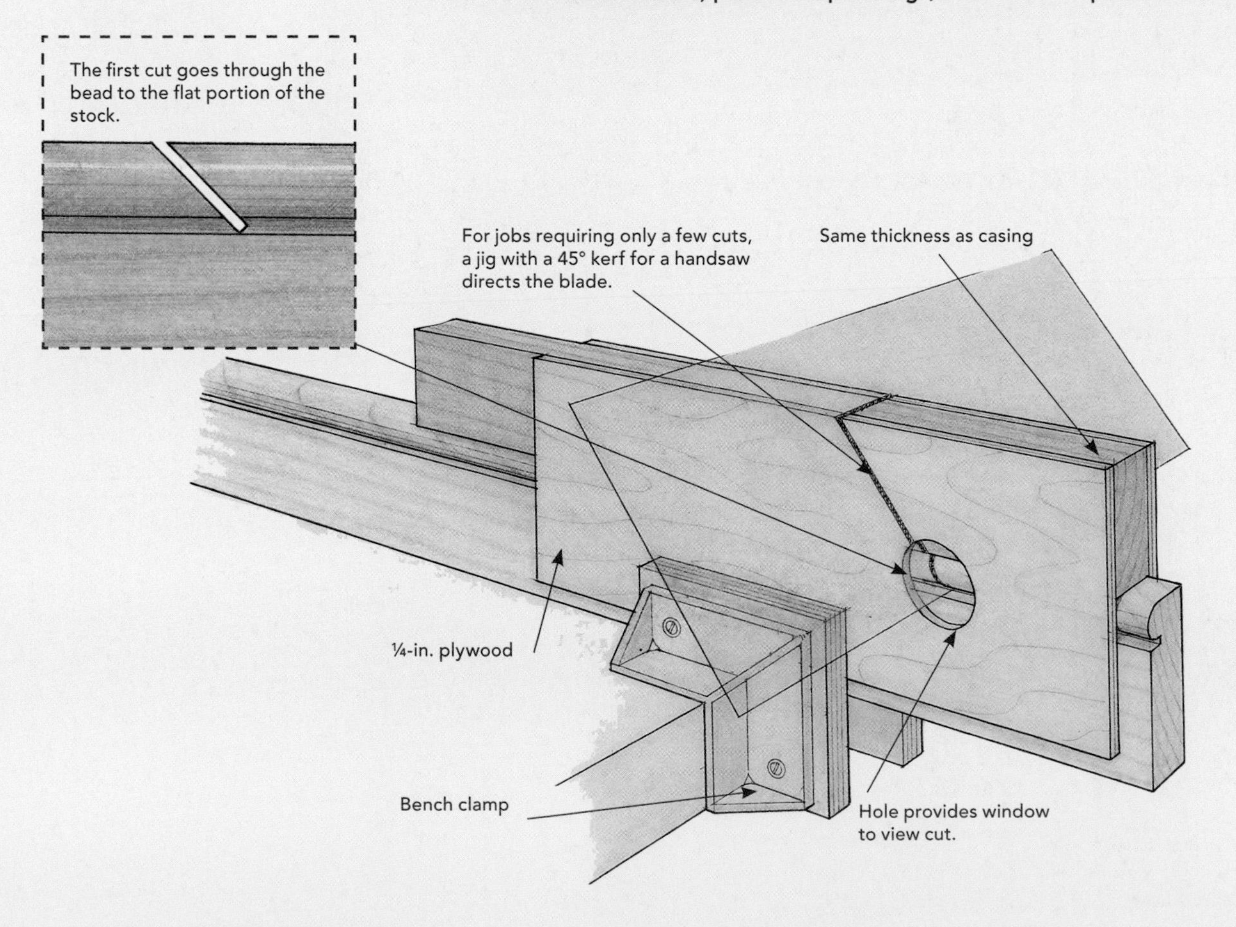

TWO TABLESAWS REDUCE SETUP TIME

A tablesaw makes the second cut along the bead at the end of the head casing; to make a similar cut on the opposite end, you have to reset the fence. Having an extra tablesaw available—one for each cut—puts an end to moving the fence back and forth between cuts.

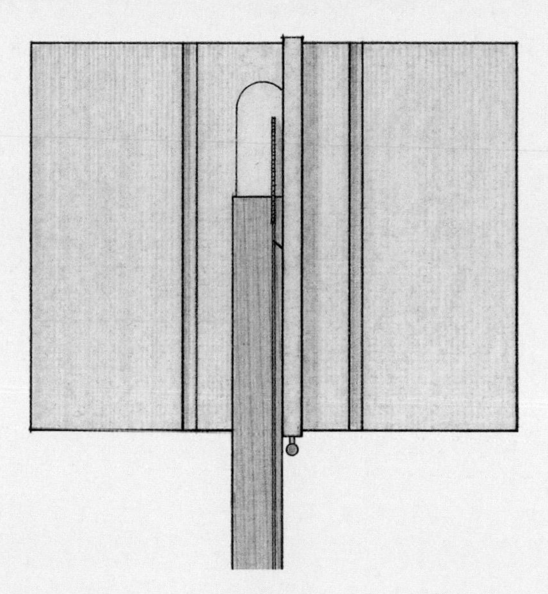

The fence is positioned so that the blade removes the bead and quirk, leaving a square shoulder on one end of the stock.

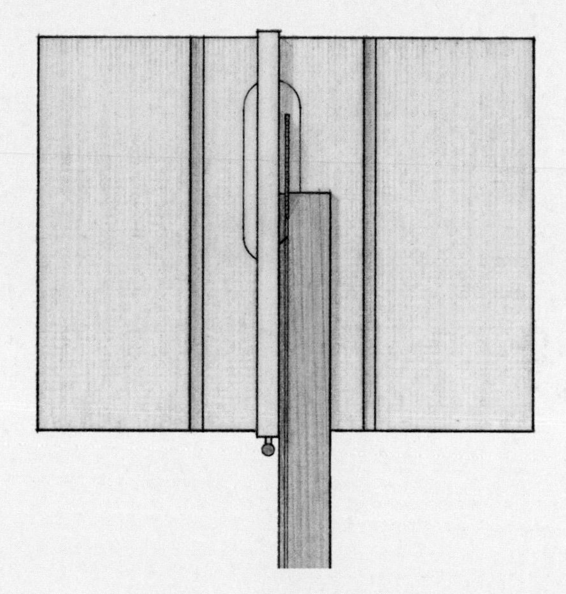

A second tablesaw setup cuts the opposite end of the head casing.

PARING is sometimes necessary to improve the joint's fit.

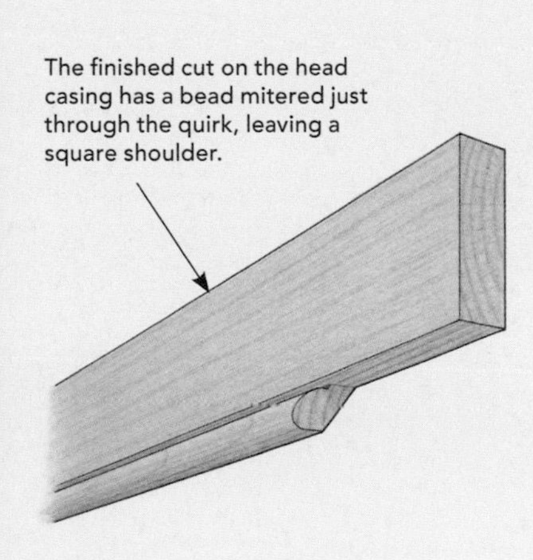

SHOPMADE JIG MITERS THE HEAD CASING

A large compound-miter saw could make this cut on narrower trim, but for the wider stock in older homes, this jig is ideal. It has three main parts: a plywood base or trough similar to an old-fashioned miter box, a pair of tracks mounted and braced above the trough, and a plywood carriage containing a circular saw, which rides in the tracks.

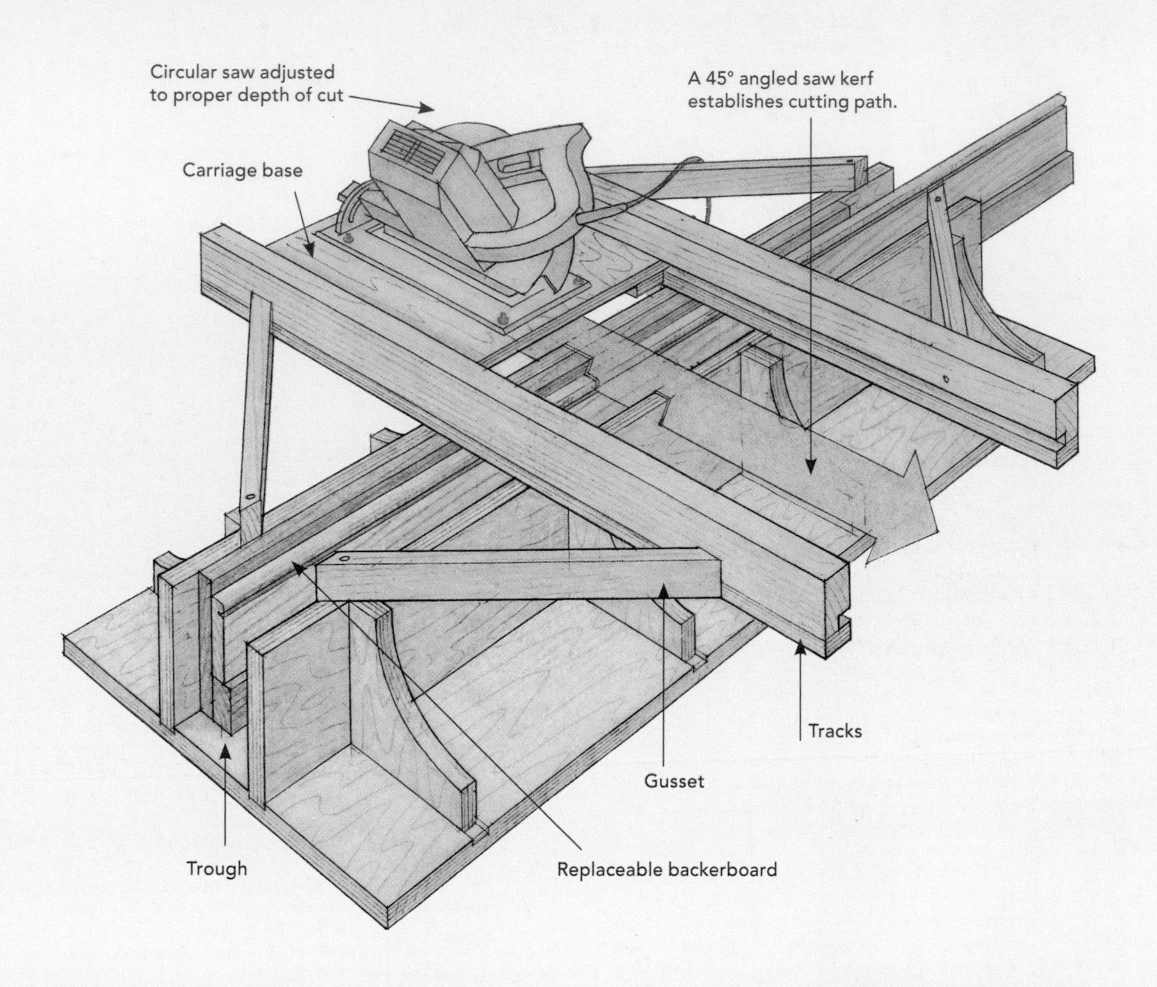

Eye screw secures stock

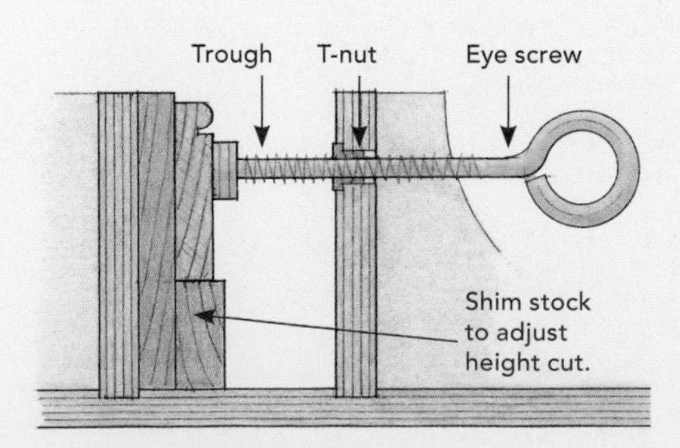

An eye screw, which is situated in the center of the vertical members forming the trough, holds beaded casing fast during the cut.

Saw rides above the casing

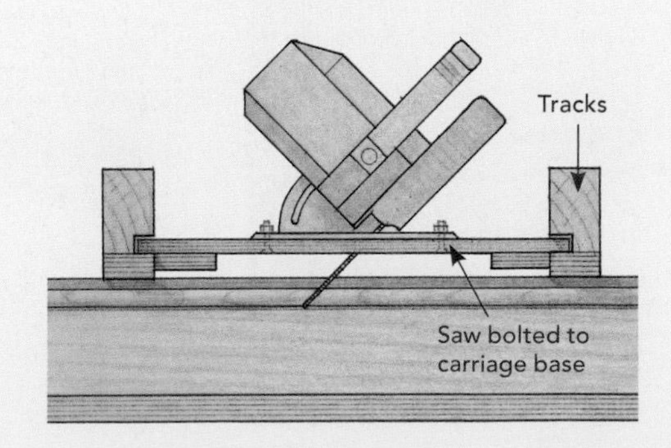

INSTALLING THE HEAD CASING

AFTER FITTING IT, mark the head casing for final length and matching biscuit slot.

USE A SLOT-CUTTING ROUTER BIT here because a biscuit joiner won't lie flat.

IF NEEDED, SHIM BEHIND THE HEAD CASING to align it with the legs.

PART 4

Paneling

Add Elegance with Raised-Panel Wainscot

BY GARY STRIEGLER

Raised-panel wainscot speaks of a time when craftsmen had an abundance of skill and the time to display their talents. But building traditional raised-panel wainscot is a complex, time-consuming process that few people can afford. I've simplified the process. Good-looking raised-panel wainscot can be built in place using basic carpentry tools and a router table. The 12-ft. by 18-ft. room pictured in this article took me three days to complete.

All the materials for making raised-panel wainscot are readily available. The bolection molding that bridges the gap between the wainscot frame and the raised panel is a stock molding from White River Hardwoods (www.mouldings.com). The raised panels are medium-density fiberboard (MDF), which will be painted. MDF profiles well, and it is more stable and less expensive than solid wood.

Start with a detailed layout

Before I cut any wood, I snap horizontal chalklines representing the top and bottom rails to see if the wainscot height is appropriate for the room. Next, I determine how many panels are needed. They should be wider than they are tall, and to my eye, an odd number of panels looks best. For this project, five panels fit perfectly on the longest wall.

I determine the panel width by subtracting the total width of the stiles (vertical pieces) from the length of the wall. One panel will overlap another in corners, and the lapped stile should be ¾ in. (the thickness of the stile) wider. Then I divide the result by the number of panels. Once the panel width is figured out, I use a level to lay out the stiles on the wall. If any electrical boxes fall on a stile or panel edge, an electrician can move them.

Build the longest frame first

Because it's easier to make fitting adjustments to the smaller frame of the shorter wall, I assemble and install the longer frame first. The frame stiles and rails must be the same thickness, so the first step is to run all frame stock through a portable planer. I also use this machine to plane all rails to width by running them through the machine on edge.

Finding clean, straight lumber in 18-ft. lengths is nearly impossible, so I use pocket screws to assemble the top and bottom rails from two shorter pieces. I make certain that the top rail's butt joint breaks on a stud and is not in the same panel as the bottom rail's joint. The overall rail lengths are ⅛ in. shorter than the wall to ensure that the rails fit easily. Any gap will be covered by the adjoining wall's overlapping stile.

I cut all the stiles on a miter saw using a stop block for accurate repetitive cuts. Next, I lay out the lumber

STOCK MOLDING AND MDF PANELS transform a room.

THE WRITING IS ON THE WALL. A full-size layout on the wall lets you see the wainscot's proportions and serves as a guide for assembly.

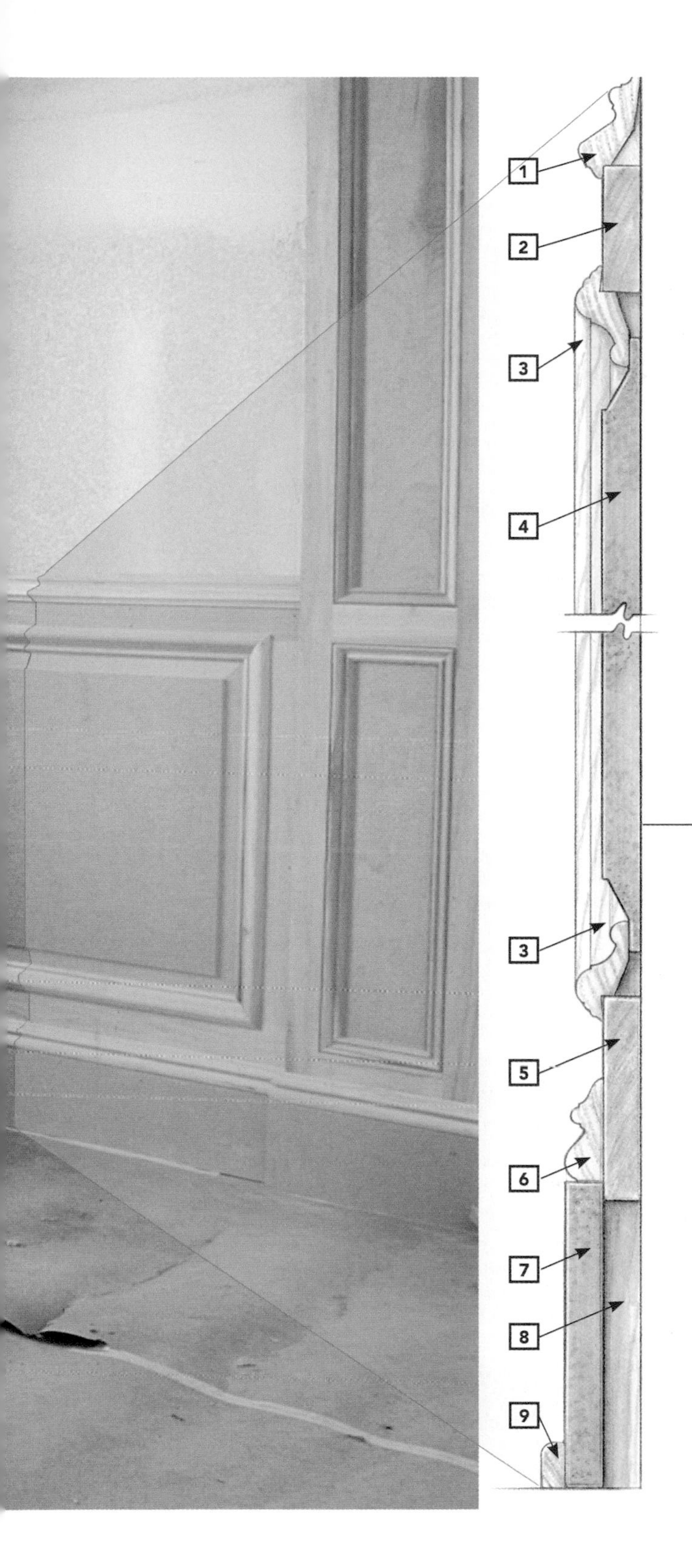

Pleasing proportions from top to bottom

Although small in section, the moldings in this wainscot have a big impact on its overall appearance.

1 Bolection-molding cap
2 Top rail
3 Bolection molding
4 MDF raised panel
5 Bottom rail
6 Panel molding
7 MDF baseboard
8 Cleat
9 Shoe molding

for the top and bottom rails against the long wall. I transfer the layout from the wall to the rails using a Speed Square. Then I lay the stiles along the rails' layout lines (see the photo above). (Remember, the lapped corner stiles are ¾ in. wider than the others.) To assemble the rails and stiles, I use a pocket-hole jig and connect the pieces with yellow glue and pocket screws (see the photos on p. 180). Glue should seep out around the joint and will be sanded smooth after it sets up.

Cleats support the frame

I nail a series of cleats along the wall to support the frame and raise it to the height of the chalkline snapped earlier. The cleats are the same depth as the frame and also act as nailers for the baseboard that is installed later.

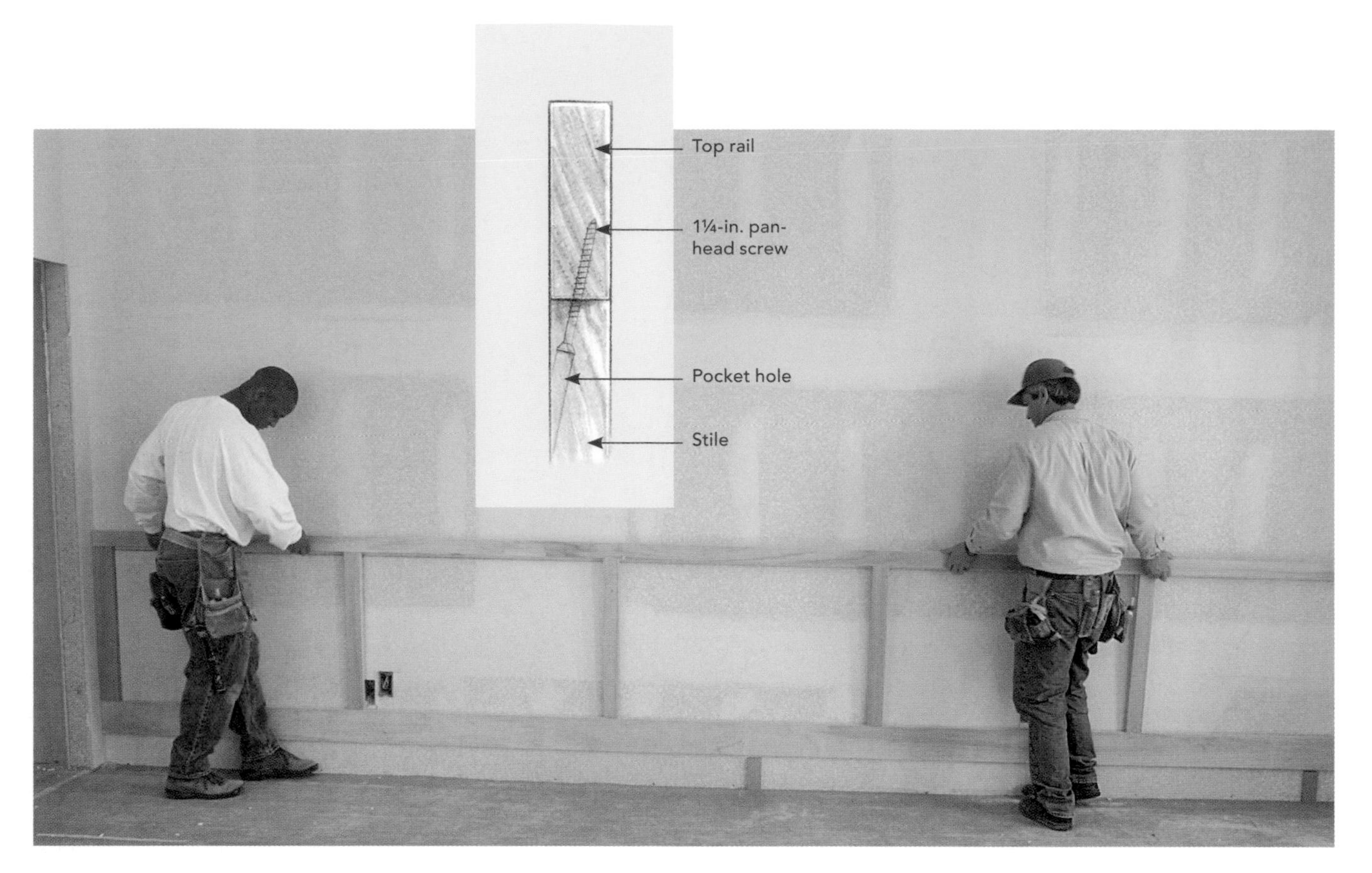

ASSEMBLE THE FRAME NEXT TO THE WALL. Use a pocket-hole jig to drill the stiles for assembly (see the left photo above). Lay out the rails and stiles along the wall, then glue and screw them together (see the center photo above). Get some help lifting the frame into place because butt joints along the rails could flex (see the right photo above). Cleats support the frame for nailing and serve as nailers for the baseboard.

Although the 18-ft.-long frame isn't that heavy, it's awkward for one person to lift into place. The pocket-screw joinery is extremely strong, but the butt joints along the rails are the weakest point of the assembly and could flex if not supported properly.

Once the frame is up, I set it on the cleats so that I can nail it to the studs using a 15-ga. finish nailer. I then smooth the rail-and-stile joints with a 120-grit sanding disk in a random-orbit sander.

Raised panels are shaped at the router table

The bolection molding for this wainscot projects 1½ in. into the frame's opening. To keep the raised-panel profile from being concealed, I size the panel to fit inside the frame and molding.

Because I knew the profile of the raised panel to be made, I determined that the bolection molding should overlap the raised panel by ⅜ in. This amount of overlap covers the thin, flat edge of the panel without hitting the raised profile, provides enough wiggle room to adjust the bolection molding, and covers the brads used to fasten the raised panels to the wall.

I cut a 1⅛-in. gauge block (the 1½-in. molding projection minus the ⅜-in. overlap) from scrap lumber and use it to lay out the raised panel in the frame. I mark only the corners because they determine both the placement and the size of the raised panels.

The rectangular panels are made from ¾-in. MDF, which I cut to size on a tablesaw. Because of the fine dust that's created when cutting MDF, a respirator is essential, even in a well-ventilated area.

The next step is to shape the raised panels on a router table. I don't rout the MDF in one pass; instead, I make three passes for each side. This process can be time-consuming, but by setting the router depth once and increasing the cut by moving the fence to reveal more of the router bit with each

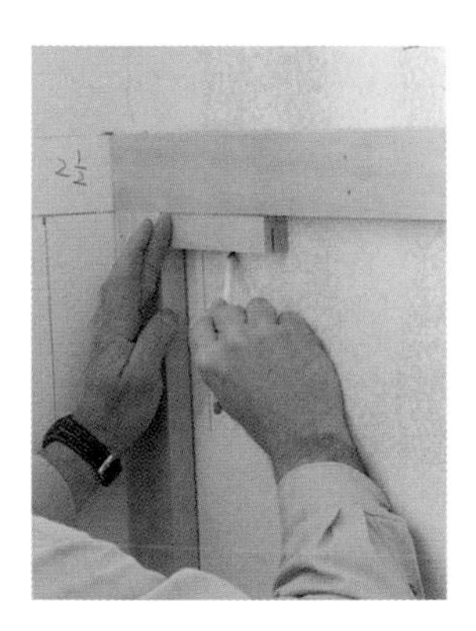

SIZE, RAISE, AND INSTALL THE PANELS. Use a gauge block to determine the panels' size and to mark their layout lines. Cutting the raised-panel profile in multiple passes on a router table ensures smooth results. Then install each panel by driving 2-in. brads through the flat panel edge.

MDF panels are flat, smooth, and stable.

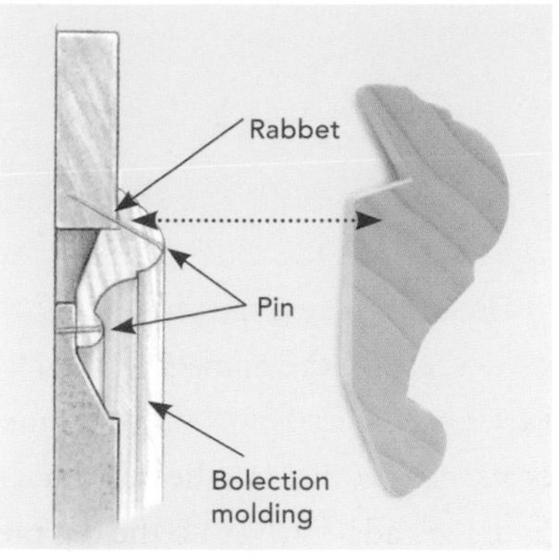

AVOID SPLITS BY DRIVING PINS INSTEAD OF NAILS. A 23-ga. headless pinner won't split even the most delicate molding. It provides sufficient holding power and makes nearly invisible fastener holes. Drive the pins into the frame members and panel edges as shown in the top right drawing.

pass, I get a smoother profile, keep the router from overheating, and reduce the chances that the panel will kick back. Plus, I don't have to duck under the table to readjust the router.

This phase is similar to a production line. I rout four sides of every panel, then increase the depth of cut, run every panel through again, and repeat the process. This approach ensures uniform panels all cut to the same depth.

To install the raised panels, I simply rest each one on a 1⅛-in.-thick gauge block along the bottom rail and position the panel along the layout lines I made earlier. Along the panel edges, I drive 2-in. brads into the studs. The MDF panels can be nailed to the wall because they are extremely stable and won't move very much with swings in humidity.

Molding bridges the gap between the raised panel and the frame

The bolection molding has a rabbet on its back, allowing it to be seated in the framed opening while covering any joints. To cut a piece of bolection molding properly, I elevate the rabbet with a sacrificial block the same depth as the rails and stiles; then it's just a matter of making 45° miters and fitting the molding to the opening.

After cutting the molding, I use a 23-ga. headless pinner to install each mitered frame by nailing the molding to the stiles, rails, and panels. I reinforce the miter joints with glue and cross-nail them to hold the joints tight.

Finish up with bolections and baseboards

Once I've finished mitering the frames around the panels, I'm in the home stretch. To cap the wainscot, I use a rabbeted bolection molding that contains a chair-rail profile (see the bottom right photo on p. 184). Brads driven into the top rail attach the molding, and the rabbet hides the joint between the cap and the wainscot. Because I use a backband on window and door casings that is thicker than the bolection molding, a simple butt joint is all that is needed to terminate the molding at windows and doors.

I make the baseboard detail out of 6-in.-wide MDF topped with a 2-in. panel molding. The baseboard is nailed to the cleats and the bottom rail, and the molding is fastened to the bottom rail. At the doorways, I clip the portion of the baseboard that stands proud of the casing with a 45° bevel to ease the transition between the two details.

A FLAT PANEL KEEPS WINDOWS CLEAN

Instead of crowding a raised panel under the window, give this area a simple step-out treatment. Use a 3½-in.-wide window stool to accommodate the extra depth for this detail. [1] After installing two MDF nailing strips equal to the thickness of the wainscot frame, build out a single, flat MDF panel 1½ in. under the window. Size it to cover the adjoining wainscot frame by ¼ in. on each side. [2] Next, install the MDF baseboard on top of the flat panel. At the step back, use a mitered return fastened with yellow carpenter's glue and 23-ga. headless pins. [3] The 2-in. panel molding that caps the baseboard should be nailed every 16 in. with 1½-in. brads.

A CAP AND BASEBOARD finish off the wainscot.

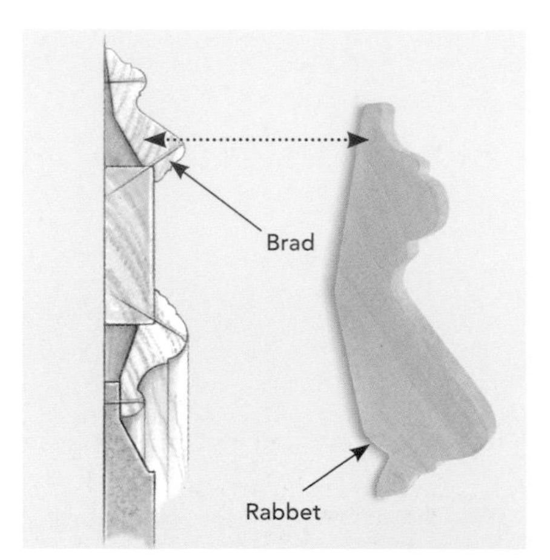

KEEP INSTALLATION SIMPLE WITH A ONE-PIECE MOLDING. Instead of a traditional two-piece chair rail, cap the wainscot with a bolection molding that has a rabbeted bottom edge. The molding contains a chair-rail profile. Complete the wainscot with 6-in.-wide MDF baseboard topped with a panel molding.

Paint-Grade Wainscot on a Curved Stairwell

BY JOE MILICIA

Working on a paint-grade trim job gives a finish carpenter the latitude to use disparate materials that blend together under the paint. The physical characteristics are what determine the choice of materials. For instance, plywood trumps solid stock when it comes to stability, a smooth surface is better than open grain, and laminations usually are preferred for bending.

On this job, one of our tasks was to run a raised-panel wainscot along the wall of a spiral staircase. Because the trim extended up to and past eye level, the layout had to be right on the money, and the finish had to be pristine. We needed materials that were flexible enough to bend easily to the curve of the wall but that wouldn't require days of finish prep.

We took over the stairwell before the stairs were measured and the drywallers had arrived. Rather than spend a day setting up staging and ladders in the two-story stairwell, I drew the layout on the bench, so to speak. Of course, it's almost impossible to use a tape measure on a curved wall, so we ripped 3-in.-wide strips of bending lauan and tacked them end to end along the wainscot's planned top-rail position to make a story pole. Once the length of the layout was established, we took down the strips and divided the total length by a suitable number of stiles and panels, then transferred these measurements back to the flexible story pole. After covering the walls with two layers of bending lauan, we nailed the story pole back in place. The stile locations were plumbed down from the top rail, and we penciled the remaining layout on the wall.

Three carpenters spent nine days in the shop and at the site to finish the installation. The completed staircase (see the photo on p. 186) is the jewel of the home's entry.

SOLID BACKING. Apply a double layer of 3/8-in. bending lauan directly to the studs; it helps to smooth out framing imperfections. As shown on the upper landing, after tacking up the first layer, make sure that the radius is fair, and shim wherever necessary. Next, apply polyurethane construction adhesive to the framing, and staple off the first layer. The second layer is applied in the same fashion; the seams should land on the framing, offset by at least one stud bay.

CURVED ASSEMBLY IS A UNION OF DISPARATE MATERIALS

The beauty of paint-grade trim is that only the surface that receives the paint has to be uniform; the substrate is chosen for its best characteristic, whether it is flexible enough to conform to curved walls or remains stable throughout the seasons. On this stair, bending plywood formed the substrate of the backing, stiles, and rails. The raised panels consisted of a hardboard and kerfed-particleboard panel picture-framed with MDF and a flexible molding.

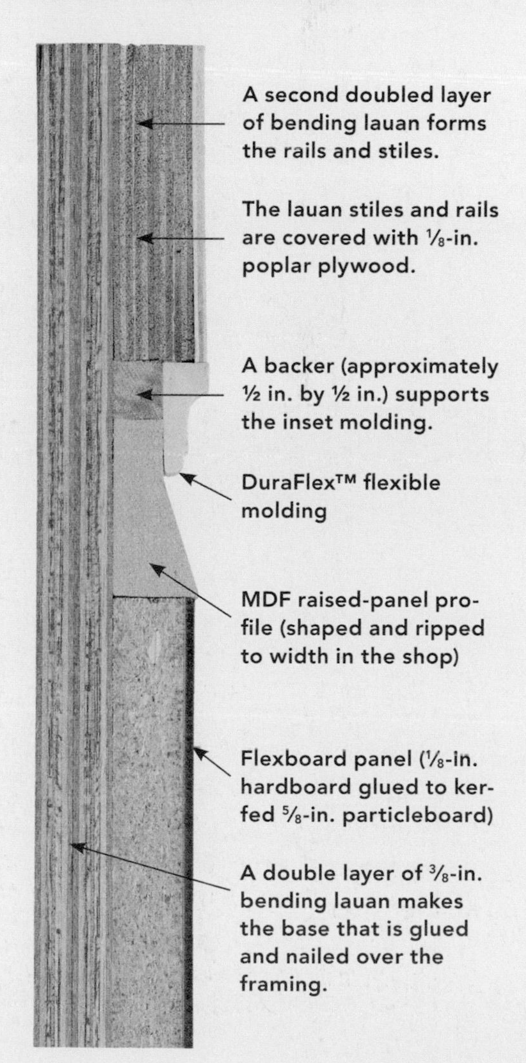

RAILS AND STILES ARE LAMINATED. Build up the rails and stiles with strips of bending lauan; cross-grain pieces work best on rails, while long grain seems to work best for stiles. Because the frame had to be ¾ in. thick, it was built up with two layers. It was important to align the layers over one another. To get consistent widths throughout, the author ripped all the pieces at the same time. Everything was glued down with construction adhesive, squared up, and stapled.

SMOOTH OUT THE FINISH WITH A VENEER. The lauan is too rough for a painted finish, so the author laminated a final layer of ⅛-in.-thick poplar plywood. Then he applied masticlike Miracle All-Purpose Adhesive (C.R. Laurence Co.; www.crlaurence.com) with a notched trowel and squeegeed the veneer into place with rubber rollers.

HERE'S WHERE SUBSTRATE ALIGNMENT IS IMPORTANT. Rip the poplar veneer wider than it needs to be. After the glue has dried, go back and run a laminate trimmer with a flush-trim bit around the veneer's perimeter. A word of caution: Any deviation in the substrate is transferred to the finish.

LIKE EVERYTHING ELSE, THE RAISED PANEL IS BUILT OUT OF PARTS. Due to the curved walls, the crew assembled the raised panels in place. The first stage was to apply spacers around the perimeter to support the molding. Next, they applied MDF bevels made by running sheet stock through a shaper fitted with a raised-panel bit, then ripping the profile from the sheet. They mitered the pieces into a picture frame that butts against the spacers.

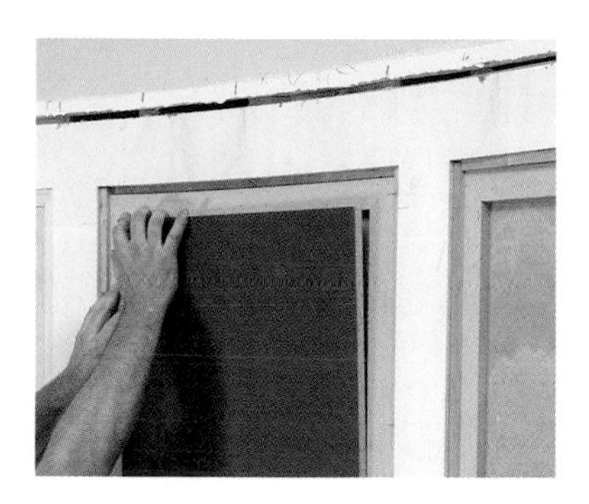

FILL IN THE CENTER. The center panel is a rectangle of hardboard backed with kerfed particleboard (Flexboard®; www.interiorproducts.com). The panel's field is measured and cut to fit tightly. Because hardboard tends to swell around nail heads, the crew used only adhesive. The last piece of the puzzle is the flexible molding (DuraFlex; www.resinart.com), which is both glued and nailed in place.

TRICK OF THE TRADE

Faced with curved work, I like to use ready-made bending plywood, often known by trade names such as Wacky Wood or Wigglewood®. Lauan, birch, or Italian-poplar veneers are laid up so that the grain orientation, rather than adding strength, allows the sheet to bend along one dimension. Available in 3⁄8-in., 1⁄4-in., and 1⁄8-in. (poplar) thicknesses, the plywood comes in 4×8 or 8×4 sheets; that is, it bends along the length of the sheet or along the width. According to wholesaler North American Plywood Corp.®, a 3⁄8-in.-thick sheet can bend to a radius of 7 in. and costs about $40; Italian poplar costs about $25 per sheet. Check your local plywood distributor for availability.

Laying Out Wainscot Paneling

BY LYNN HOPKINS

Historically, most paneling in houses was wood. It was available, it was beautiful, and it was more durable than plaster. But solid wood has a pesky attribute: It moves as temperature and humidity change.

Our clever forebears figured out how to turn this limitation into an advantage. They floated panels in a rail-and-stile frame. The panels were beveled around the perimeter so that they were thick enough in the center not to warp, but thin enough at the edges to fit into the grooves or rabbets in the rails (horizontal pieces) and stiles (vertical pieces). Known as a raised panel, this design not only allows for wood's natural movement but also creates attractive profiles and shadowlines.

Around the turn of the 20th century, the advent of sheet goods meant that a thinner, flat, dimensionally stable panel could be used. In flat-panel construction, the rails and stiles are still grooved or rabbeted, but seasonal movement isn't a concern.

Composite paneling is a flat-panel system embellished with molding, or sticking, applied where the panel meets the rails and stiles.

History can help guide your choice between raised and flat panel styles. Rooms that strive to achieve a colonial, French provincial, Victorian, or other period character predating the late-19th century are appropriately clothed in raised or composite paneling. Rooms that take their cues from the simpler styles of the mid-19th century on, including the Shaker, Craftsman, deco, and modern eras, will feel most at home with flat panels.

Wainscot paneling covers the wall to the height of a chair rail, typically 30 in. to 42 in. above the floor. It is a popular height for kitchens, breakfast areas, and dining rooms, where the paneling serves a protective as well as a decorative role. Taller paneling is appropriate in more-lavish or more-intimate rooms (see the drawing on p. 195).

The drawings on the next several pages demonstrate how I handle some of the common issues that arise when laying out paneling in a room.

Three steps to a perfect layout

Start with a scale drawing of each wall in the room, showing all doors, windows, electrical outlets and switches, air vents, radiators, and other features. The baseboard and door and window casings should also be shown on the drawing. The baseboard is usually the widest trim, followed by door and window casings.

The first step to laying out the paneling is to set the chair rail of the wainscot at the desired height. Wainscoting is typically between 30 in. and 42 in. tall. I like to locate the chair rail one-third to

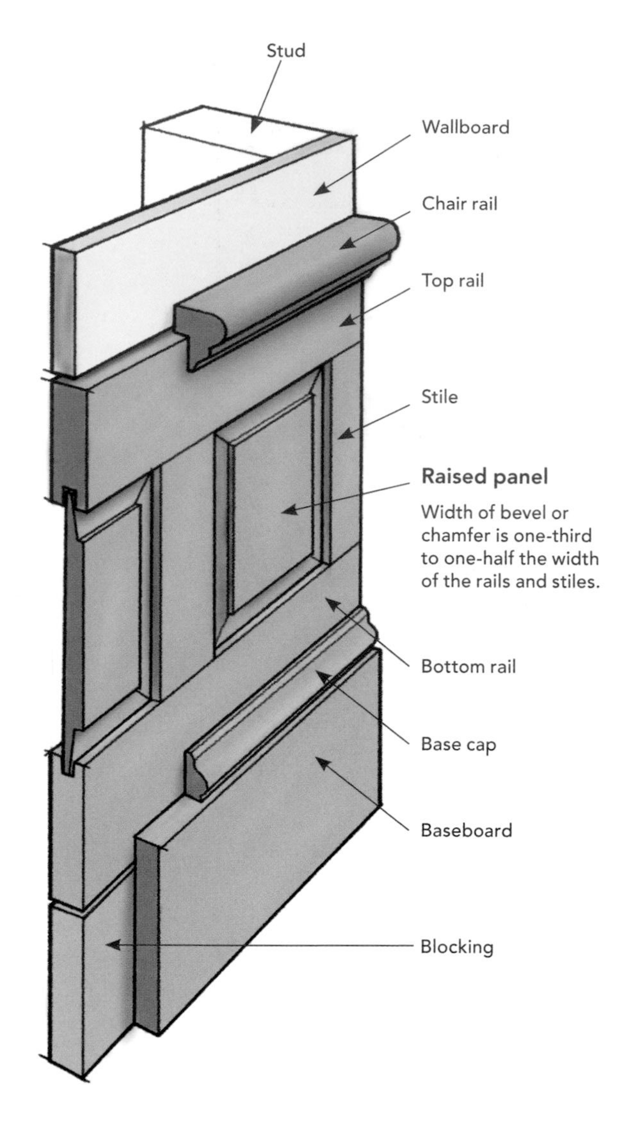

ANATOMY OF A PANEL

Raised panels are the most traditional and formal style of paneling. Their elegant form derives from the practical need to prevent unsightly gaps from appearing as solid-wood parts expand and contract.

Raised panels are the most expensive to make and install. They require material that can be beveled, such as solid wood or medium-density fiberboard (MDF). Panels must be measured, cut, and beveled to close tolerances.

Dimensionally stable flat panels are less expensive to fabricate and install than raised panels.

Composite paneling relies on sticking for its profile. Consequently, it can be easily field-adjusted like a flat panel, yet provide the visual interest of a raised panel.

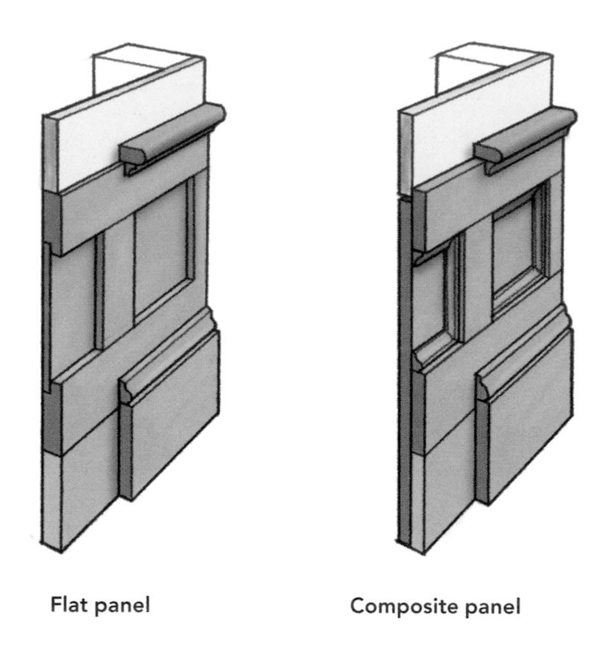

two-fifths of the way up the wall. I use a molding 1½ in. to 2½ in. wide.

The second step is to draw the rails and end stiles. Once I've determined the height of the chair rail, I draw the rails. The proportions of the rails and stiles should relate to the other trim in the room. The rails should be between one-third and two-thirds the width of the baseboard. I usually make my rails and stiles between 2 in. and 3½ in. wide and keep the stiles the same width as the rails. Draw an end stile next to the side casings of each window and each door.

The final step is to add the intermediate stiles that create the frames for the panels (the blue-gray shaded

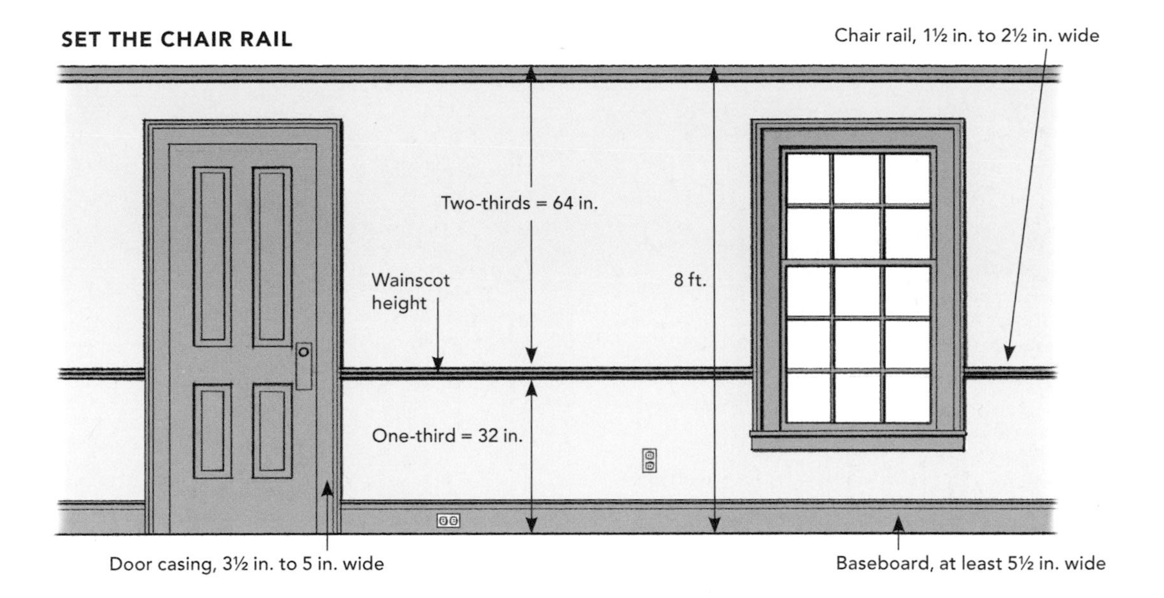

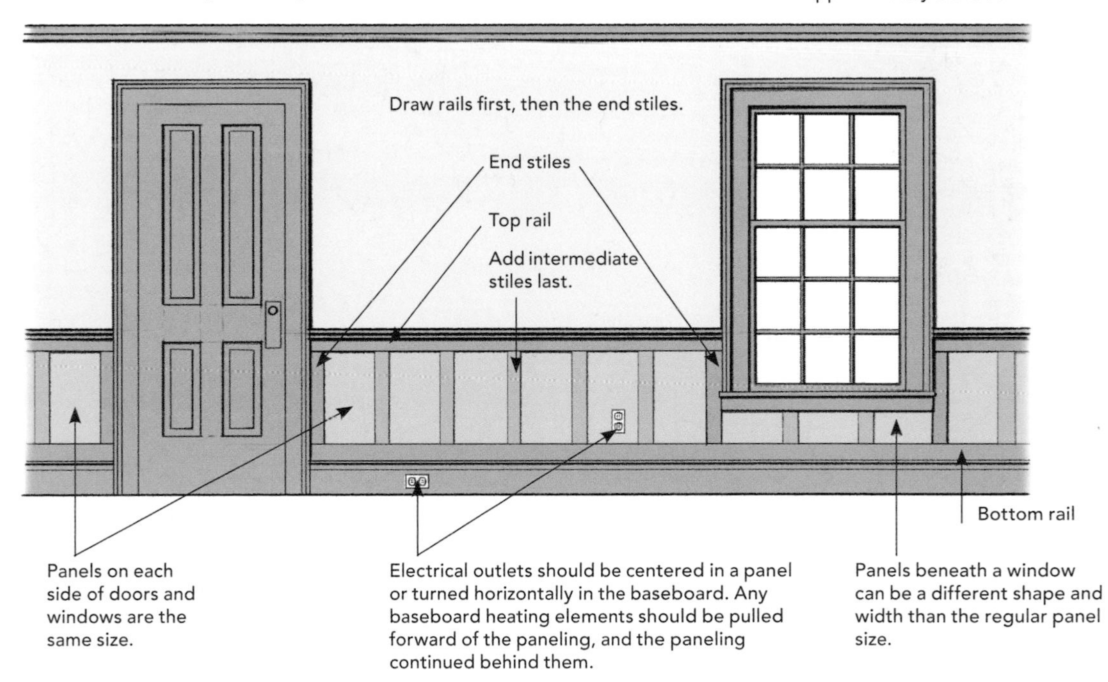

stiles in the drawing above). Paneling will be most pleasing if you establish a regular module that repeats as it wraps around the room. Sometimes this is easier to accomplish with wide, horizontal panels. Other times, narrow vertical panels look best. Here's where the artistry of trial and error comes in.

In sizing panels, I often aim to create rectangles with a 3:5 ratio. It's smart to begin in the center of the wall or in the center between two windows or doors, and to work toward the corners. The space left over at the end of the section probably won't be a typical panel width. (The end panel can differ from the standard module.) Once you have a panel

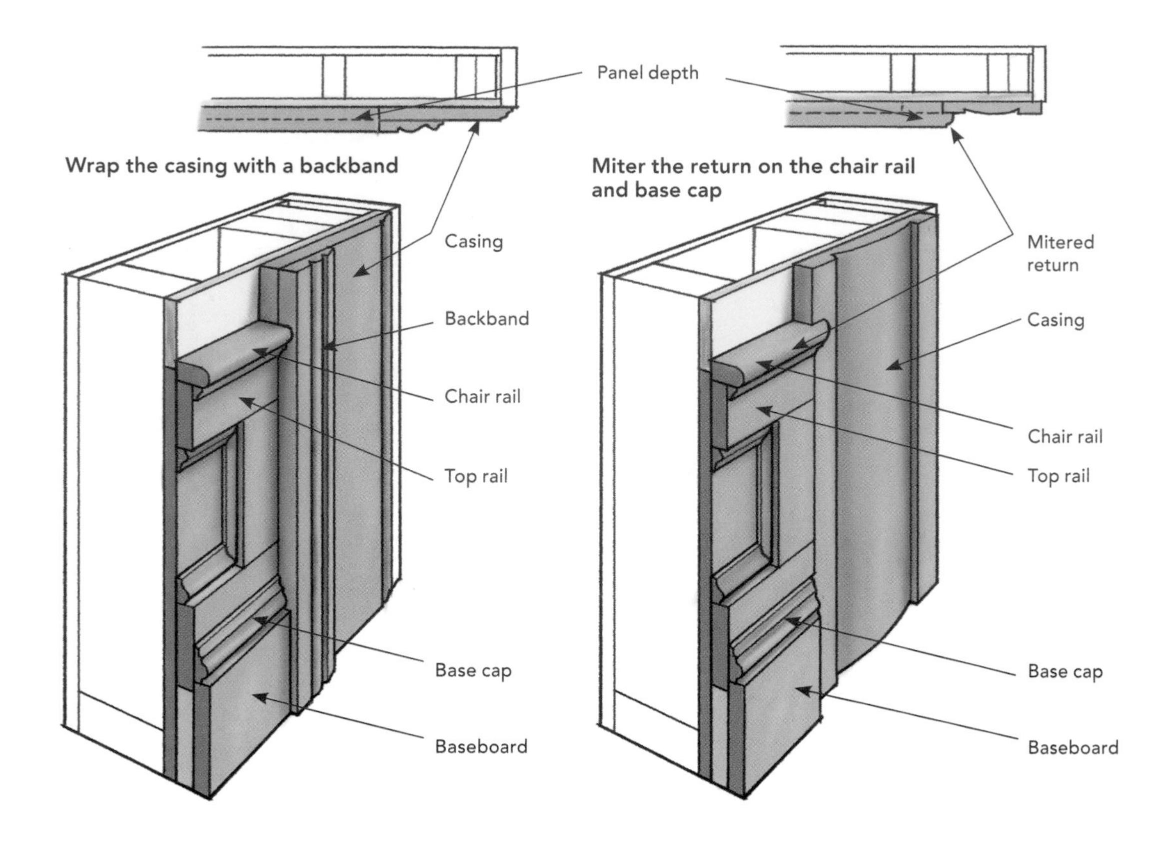

module, take that width to the other walls and chase it around the room. You might find that you need to adjust panel widths as you lay out the room. It could take a few trials before you find the standard panel width that creates the most-pleasing corner panels, but remember that it is better to work through these issues on paper than in wood.

Whether you place stiles in the corners depends on the circumstances of each room, but be consistent. Either each inside corner should have two stiles (larger end panels), or none of the corners should have any (smaller end panels). Generally, it is best to use stiles on each side of all outside corners.

The horizontal panels below the windows will be their own unique width, although you might wish to subdivide them.

Casing should be thicker than paneling

As you plan a room, consider how the paneling meets the door and window casing. If the outside edge of the casing stands proud of the paneling's top cap (or chair rail), base cap, and baseboard, then life is easy. You can butt the paneling against the trim. If not, then you need to deal with the different material thicknesses.

Although a mitered return on the top cap, base cap, and baseboard is a frequently used solution (see the top right drawing), I think it is cleaner and easier to use casings thick enough to avoid a mitered return. One solution is to wrap the casings with a thicker backband (see the top left drawing). If the casing has a flat profile, you might be able to use the top-cap profile as the backband.

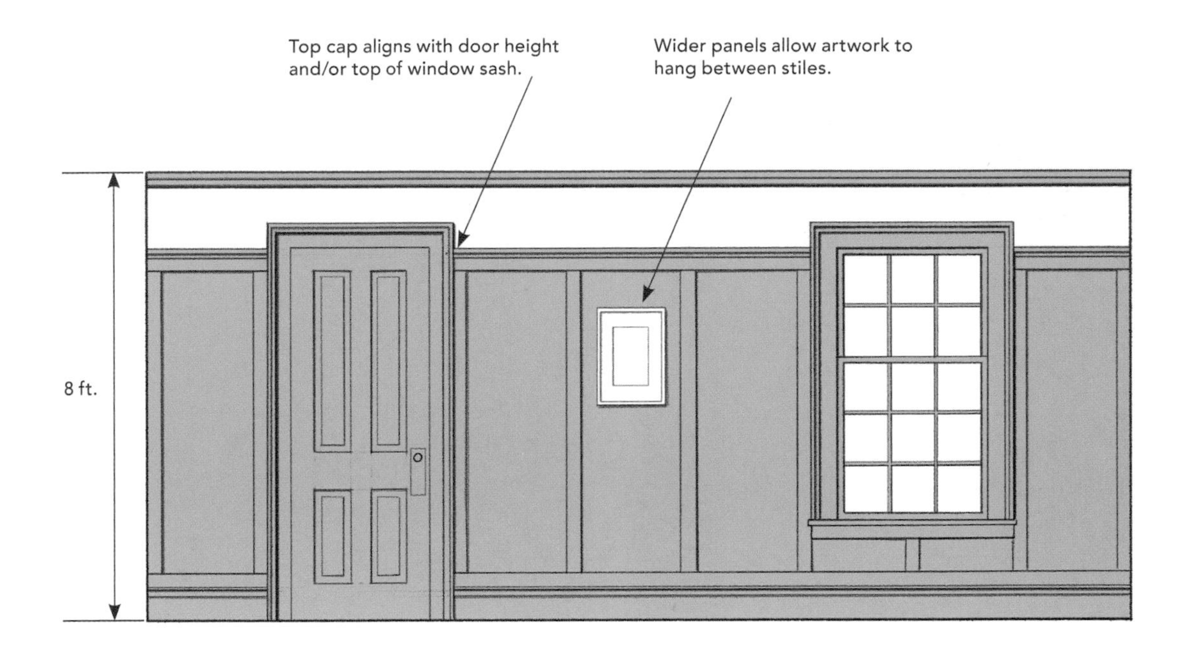

Use taller paneling in formal or intimate rooms

Because there is little practical reason to protect the upper wall, taller paneling is more lavish than wainscot; the extra height is primarily decorative. This more-regal height is an appropriate choice for living rooms, libraries, and some dining rooms. Paneling that rises to shoulder height or above provides a feeling of enclosure, protection, and intimacy, making it comforting for bedrooms. It's also good in rooms with high or cathedral ceilings where wainscot height would seem out of scale.

Taller-than-wainscot paneling that doesn't go all the way to the ceiling has two standard heights. Plate-rail height is about shoulder level and is worth considering if you have plates or something similar in size you wish to display at eye level. But this height can be restricting because hanging artwork above or below the rail puts these items at an awkward viewing level.

Picture-rail-height paneling is a more-flexible option. I typically set the top cap at the height of the door and/or the top of the window sash. The casings around the doors and windows will be higher than the paneling cap. I prefer this design as it acknowledges that windows and doors are important enough to interrupt the paneling.

Wider spacing of the stiles with taller paneling makes it possible to hang artwork at a comfortable viewing height within the panels. The wider the spacing, the wider the art you can accommodate without straddling a stile. An odd number of panels between doors and windows will let you center artwork in a panel and between the doors and windows. With wide spacing between the stiles, it often looks best to use two stiles in each corner of the room.

Designing with Wainscot

BY JOHN S. CROWLEY

What makes old houses so evocative? Is it an implied promise of longevity? Maybe it's our romantic associations with details like window seats, big fireplaces, and architectural woodwork. One of an old house's more attractive details is wood wainscot, which imparts special warmth to a room. As a part of a historical continuum, this centuries-old feature can connect even the newest home to a more attractive past.

The building boom of the past few years has awakened interest in historic styles that make a house seem more like a home and less like a box with windows. It makes sense for builders and designers to refer to the past. Happily, the past 500 years have left an indelible record of wainscot variations that exemplify particular styles, from colonial to modern.

From insulation to decoration

In northern Europe in the 15th century, wide boards or panels were first used to draftproof rooms. These wainscot panels evolved to fit in a modular framework that covered an entire wall surface. As architectural styles grew more complex, wainscot followed suit and soon featured ornate carvings, motifs, family symbols, geometric shapes, and elaborate paint finishes. At times, wainscot was considered a valuable heirloom to be taken with a family from one house to another.

In the New World, architectural design was driven first by immediate concern for shelter and by a stiff spirituality that rejected any ornamentation. As America grew into the 18th and 19th centuries, fancier designs became one hallmark of a burgeoning economy. Victorian Americans carried ornament to its most complex, and like their European ancestors, they were succeeded by the reform-minded. This time, the reform was less about religion; the Arts and Crafts movement pared design back to more essential ingredients. Later, the 20th-Century Modern movement led by Louis Kahn and Alvar Aalto sought a similar appearance, a minimalist, smooth, and unarticulated wall surface.

In the next pages, I'll describe the three major American wainscot styles and discuss a few points of design.

GETTING THE PROPORTIONS RIGHT

IN DESIGNING CLASSICAL WAINSCOT, the most important concept is tripartition, which divides a wall into three fundamental parts: the wainscot, the field, and the frieze (see the drawing at right). The wainscot may vary in height from chair rail to ceiling (eliminating the field and frieze when it does so).

Increasing the wainscot to a height two-thirds of the distance from floor to ceiling creates a strong visual design statement and the feeling of being surrounded by the warmth of wood. This height is particularly popular in Mission, Craftsman, and Arts and Craft designs. Even a full-wall Colonial-Revival treatment shows evidence of a tripartition design (see the photo on p. 197).

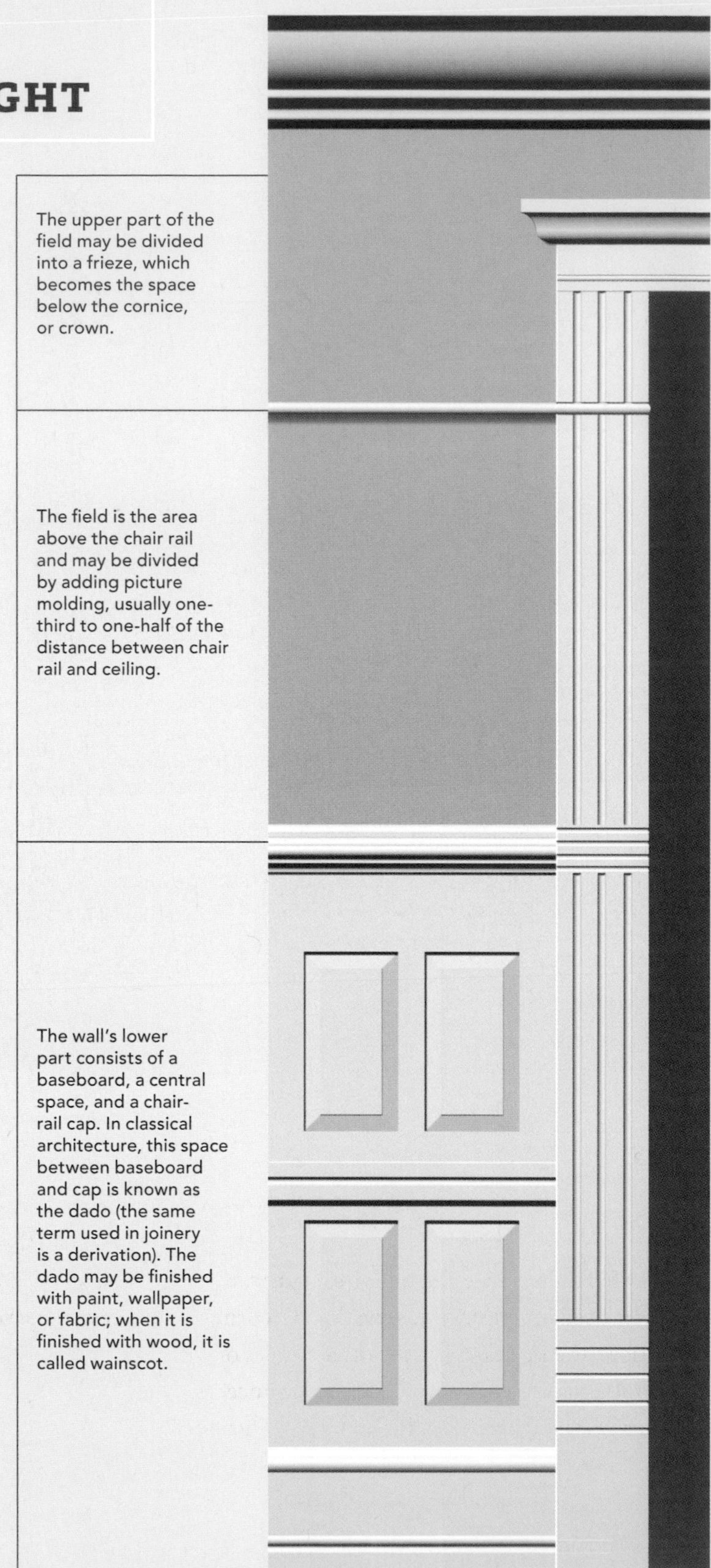

COLONIAL RESERVED DECORATIVE

Based on classical design principles, a typical example consists of beveled panels captured by a framework of rails and stiles.

COLONIAL HIGH DECORATIVE

Used in the more formal public areas of a house, an ornate style often was expanded to a floor-to-ceiling treatment of the fireplace wall.

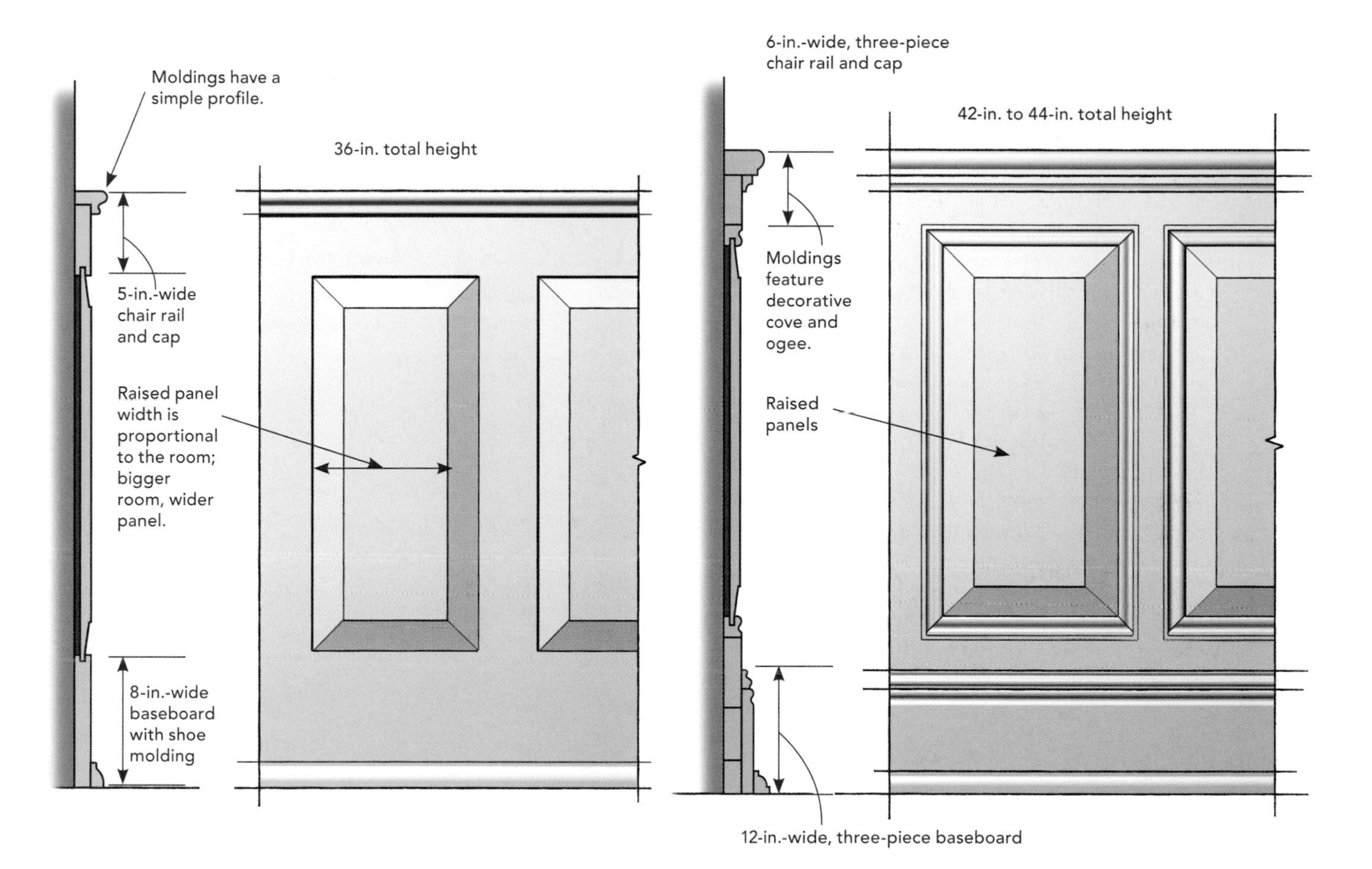

The Colonial style began with simplicity

Brought from Europe to the colonies, wainscot often was used for energy conservation, not ornamentation. The Puritans saw the ornate styles of their motherland as an unwanted extravagance. As America grew during the 18th century, the Puritans' grip on style loosened, and paneling styles began to reflect a diverse mix of cultures that included more classical motifs. The popular Colonial style has been revived periodically for 200 years.

VICTORIAN RESERVED DECORATIVE

Keeping intact the idea of the raised panel and frame, Victorian wainscot often featured intricately detailed moldings.

VICTORIAN HIGH DECORATIVE

Designers often raised the decoration by increasing the complexity of the panel arrangement and by using figured hardwoods.

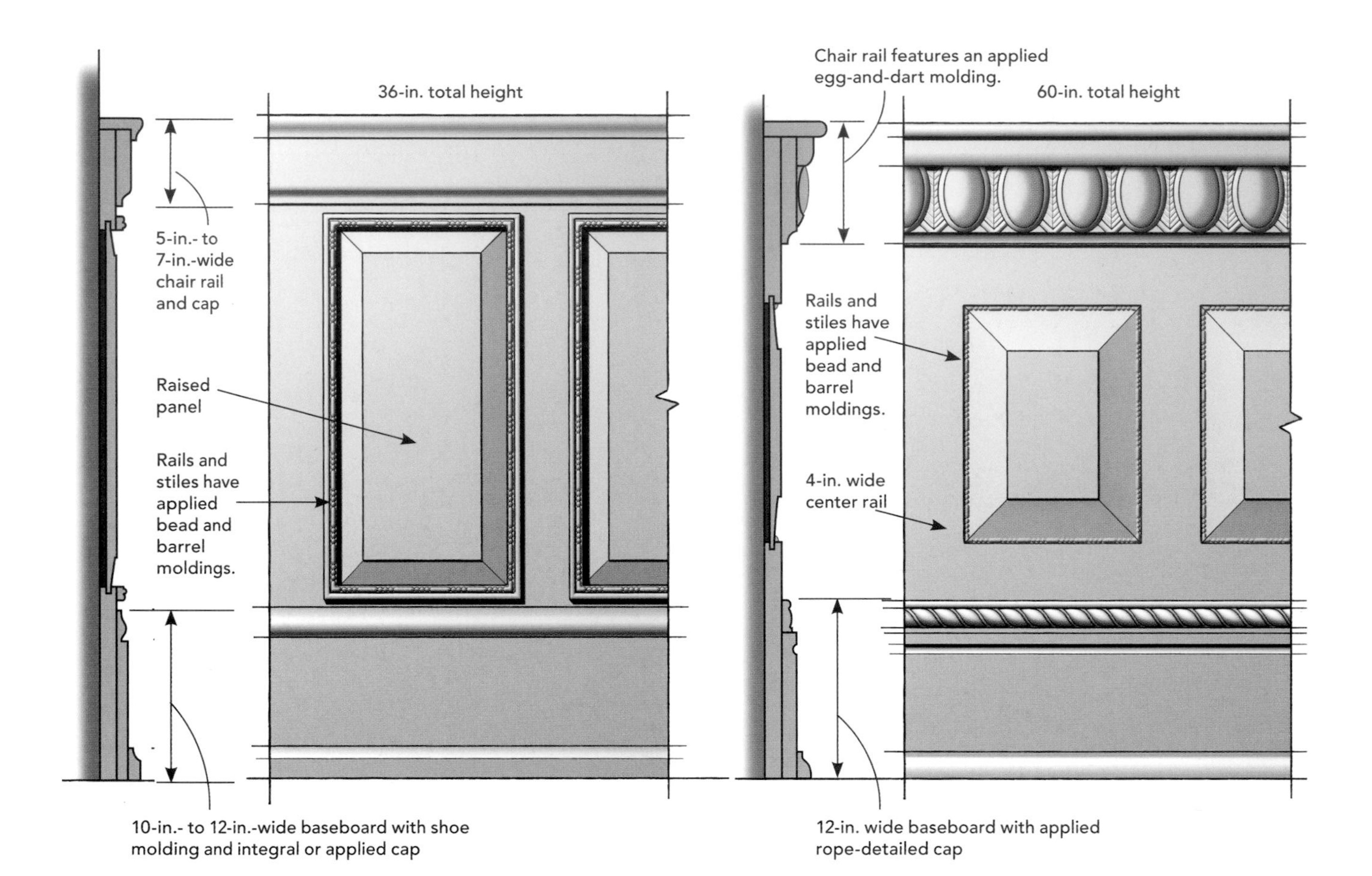

The rise and fall of Victorian ornamentation

The stylistic diversity begun in colonial times only increased during the Victorian era. The advent of a wide variety of machine-made millwork that was sold by catalog gave designers the material to embellish any of the era's styles: Empire, Stick, Queen Anne, Shingle, or Gothic.

It was also during this period that the restrained beadboard emerged. Commonly used in kitchens and pantries, beadboard became the cost-effective choice for summer cottages and eventually was associated with an informal lifestyle.

VICTORIAN FOLK STYLE

Occasionally paired with panels above, typical beadboard was installed only to chair-rail height.

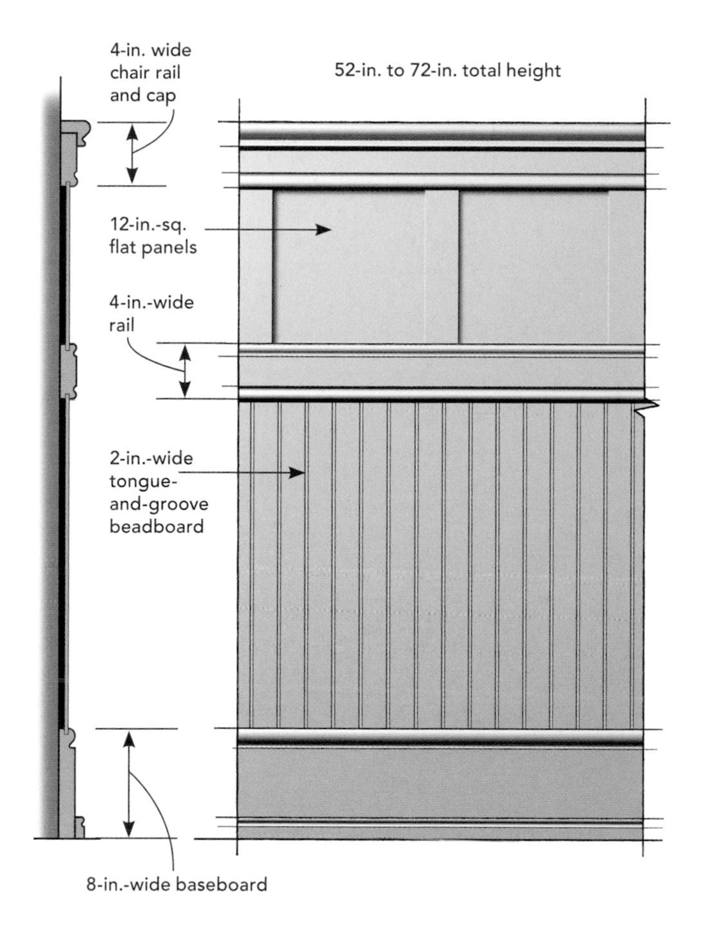

20TH-CENTURY PRAIRIE

Featured in custom-designed homes, the simple lines of this wainscot often were enhanced with the use of clear finishes.

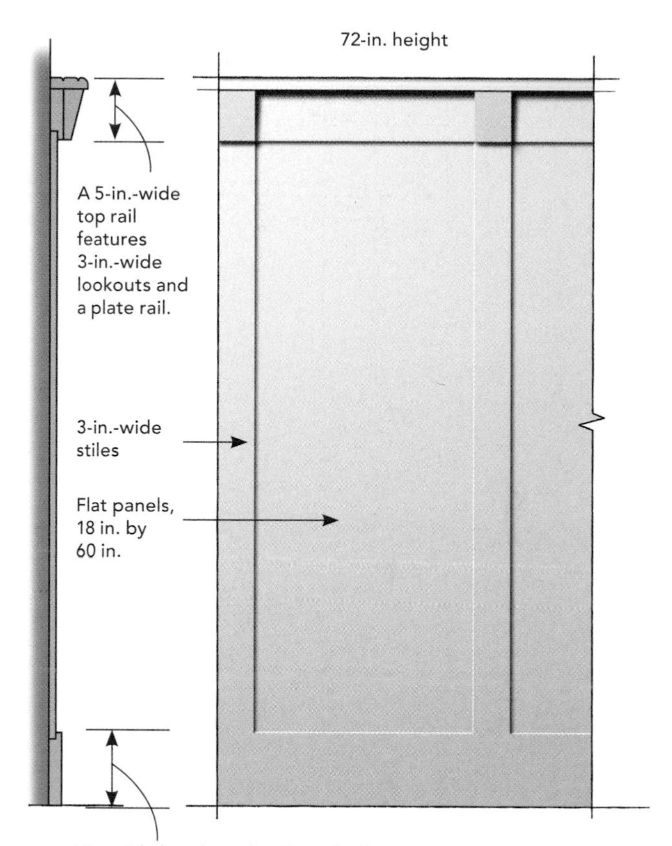

20th-century styles simplify

The Craftsman movement was an early 20th-century response to the Victorians' excesses. As the name suggests, Craftsman—along with Mission and Prairie styles—popularized a minimalist beauty that was based on structural integrity and fine joinery. Wainscot design featured flat panels and simply detailed frames.

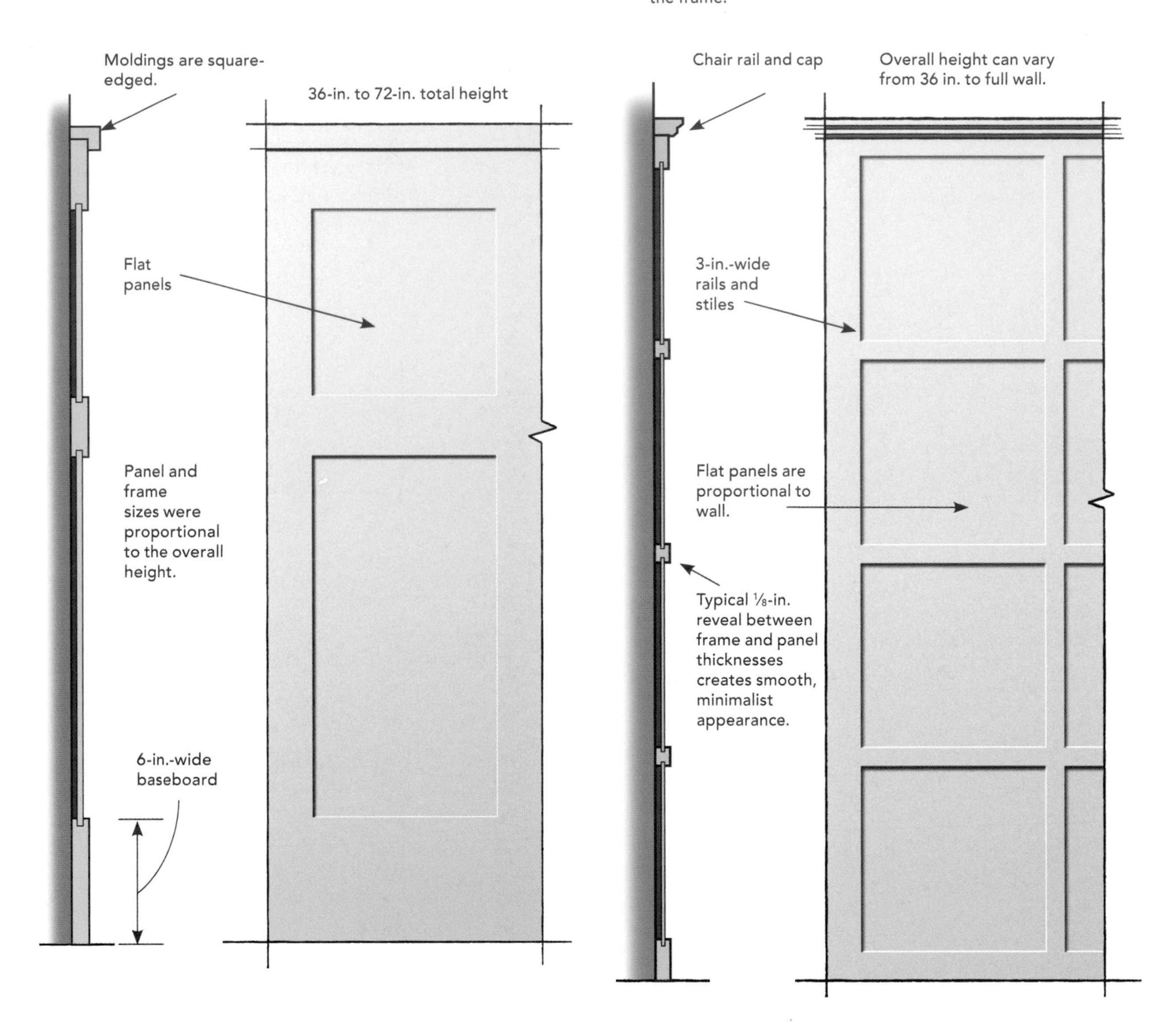

More Information

To find out more about architectural styles and history, you can start by contacting your local historical society. Museums are also a great resource; the listing below is just a tiny portion of the archives and exhibitions that are available to the public.

SOCIETY FOR THE PRESERVATION OF NEW ENGLAND ANTIQUITIES
www.spnea.org

THE WINTERTHUR MUSEUM
www.winterthur.org

THE HILL-STEAD MUSEUM
www.hillstead.org

THE METROPOLITAN MUSEUM OF ART
www.metmuseum.org

THE NATIONAL BUILDING MUSEUM
www.nbm.org

GENERAL CONSTRUCTION DETAILS OF WAINSCOTING

Inside corners. Butt joints are preferable for wider plain stock, but smaller ornate moldings can be mitered or coped.

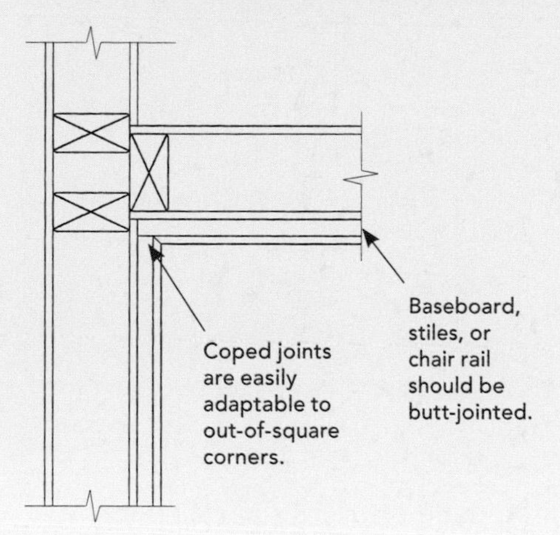

Outside corners. To minimize the effects of seasonal movement, outside corners on stiles should be made with a half-lap or rabbet joint, not a miter. Smaller moldings such as caps and shoes can be mitered.

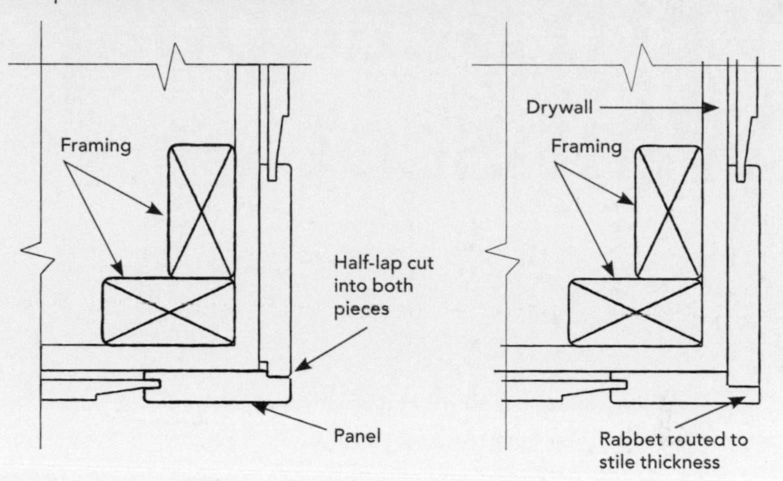

Intersection of chair rail and casing. The options for handling the transition from the chair-rail cap to a vertical casing are determined partially by the finish. Exposed end grain is typically acceptable only when painted.

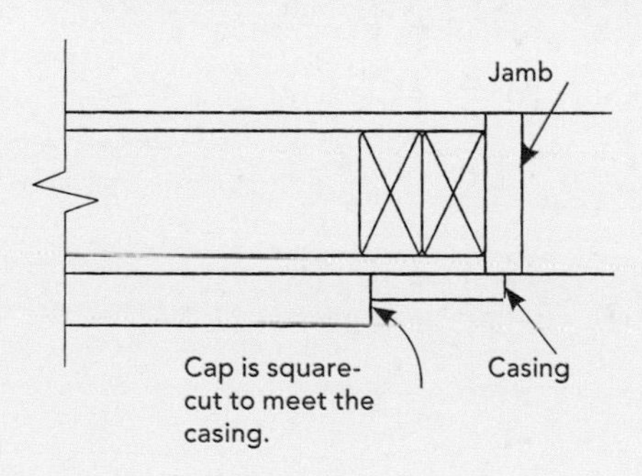

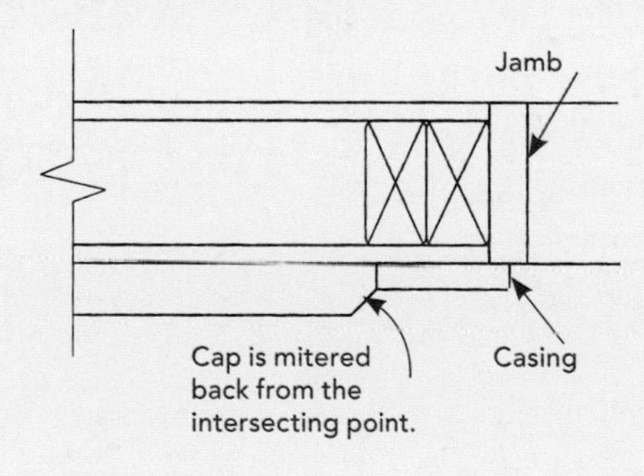

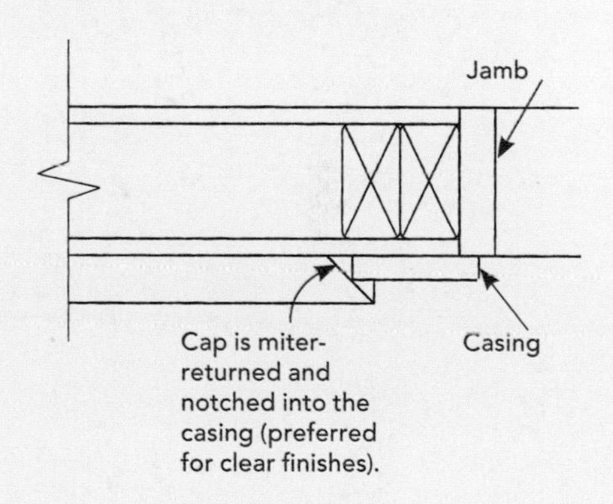

Outlets and phone jacks. In new construction, outlet or switch boxes should be located so that they fall in the center of a panel, rail, or stile. During a renovation, if the box can't be moved, build a frame around the box, install the wainscot around it, and extend the box. This technique also works for heating units and vents.

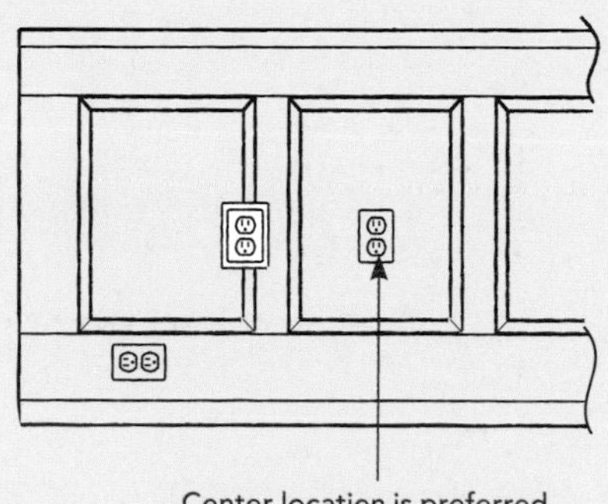

Mounting frame is slightly larger than the cover plate.

Add Style with Beadboard Wainscot

BY RICK ARNOLD

I love to transform a room with wainscot. My favorite type (and the easiest to install) is beadboard (see the sidebar on pp. 210–211). Whether in a Victorian house with stained beadboard panels or in a cottage with painted beadboard, a room can morph from blah to beautiful quickly. By matching trim details such as the cap (chair rail) and the baseboard with the rest of the millwork (casing, crown), wainscot can enhance the whole style of a room.

For this project, I installed paint-grade tongue-and-groove beadboard planks, which are the best type I've worked with. These ½-in.-thick preprimed boards, made from finger-jointed stock, have a finished width of about 3 in. After the first board is secured, the rest are mostly self-supporting, so whole walls can be dry-fit. Because baseboard hides the bottom ends of the boards in most cases, installation goes quickly, and I have to pay attention only to critical areas.

IF YOU KNOW HOW TO HANDLE A FEW TRICKY SPOTS, this elegant detail can be a snap to install.

FORGIVING INSTALLATION

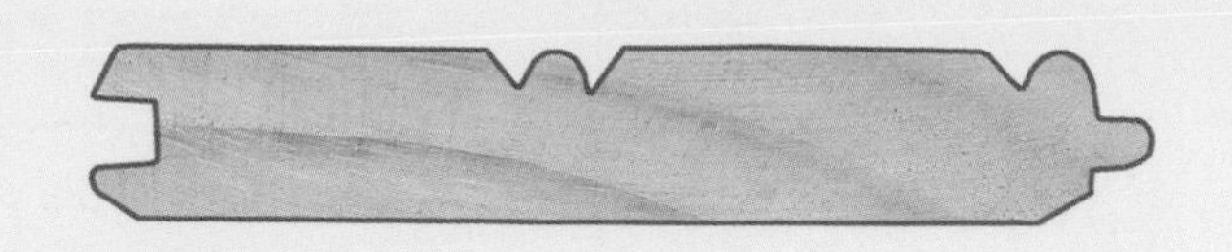

To complement the colonial trim in this house, the wainscot cap is a simple bull-nose, which rests on top of the beadboard and is supported by a ¾-in. cove molding. A standard 3½-in. colonial base covers the bottom edges of the beadboard and returns at the edge of the casing.

FINGER-JOINTED BEADBOARD

For paint-grade beadboard, finger-jointed wood has all the benefits of both regular wood and medium-density fiberboard (MDF) stock. It has the feel, workability, and look of standard lumber, but it enjoys the stability of MDF. Also, its dimensional measurements stay consistent from one board to the next. On this project, I used WindsorOne SPBC4 beadboard (www.windsorone.com).

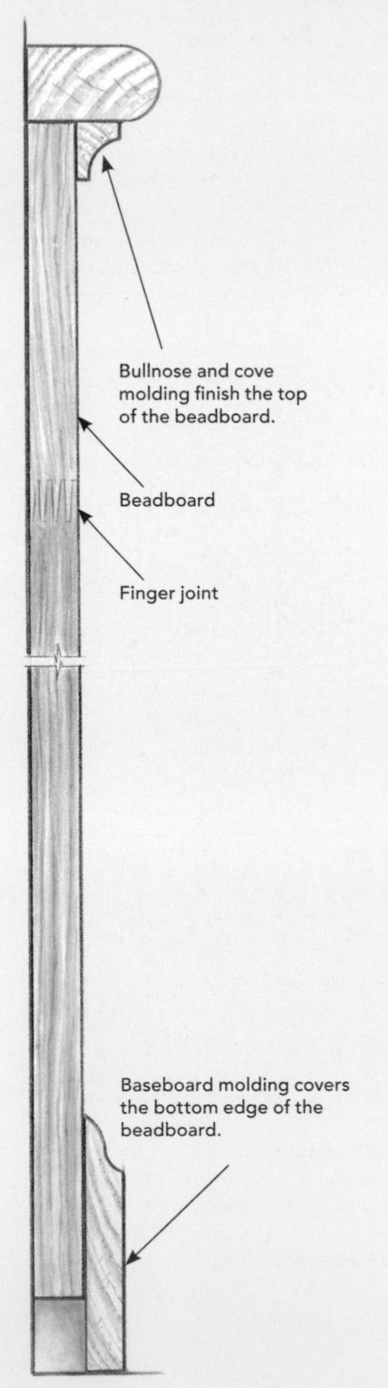

Beadboard basics

I like to glue beadboard to a clean substrate, so with existing drywall, I cut an inch or so below the top of the future wainscot cap and remove the drywall from there down. In older houses where removing lath and plaster would be messy and impractical, I clean the area as best as possible and make sure to attach the beadboard to the plaster with polyurethane glue.

TIP: Make clear marks on the plywood to pinpoint nailing hazards behind the wall, such as plumbing feeds or wires.

A GUIDE STRIP KEEPS THE TOP STRAIGHT. Start by snapping a level chalkline around the room to indicate the top of the beadboard (usually about 35 in. from the floor). Then tack on a temporary straightedge with small brads. Keeping the top of the beadboard in a straight line provides an even edge for the cap to rest on.

START PLUMB AND STAY PLUMB. Tack the first board in place, and check it with a level to make sure it's plumb. If need be, plane the board on the casing side until the leading edge (the one with the tongue) is plumb. Then check every couple of feet to make sure that the beadboard is staying plumb. You can correct for plumb in the field by opening a slight gap between two boards at the top or bottom. (If a correction of ⅛ in. or more is required, spread it out over a few boards.)

NAIL WHERE IT WON'T SHOW. In place of nailing along the length of the boards, run horizontal beads of construction adhesive about 16 in. apart on the plywood. Then face-nail the top and bottom of each board where trim will hide the nails.

Fit around windows basics

Before running the beadboard, I install the casing and the stool (without returns) around each window (see the bottom left photo). But I leave off the apron until after the beadboard is installed.

The ability to dry-fit tongue-and-groove beadboard is a real plus when installing it around a window. To get precise cutlines around window casing and other obstacles, I make a scribe guide by ripping the tongue off a length of beadboard. This gives me the exact finished width of a board. Each side of the window receives slightly different treatment depending on the direction I run the beadboard.

APPROACH SIDE OF THE WINDOW

SCRIBE THE VERTICAL LINE... First, dry-fit the last full board before the window casing. Then, placing the scribe guide tight against the casing, trace the cutline on the dry-fit board.

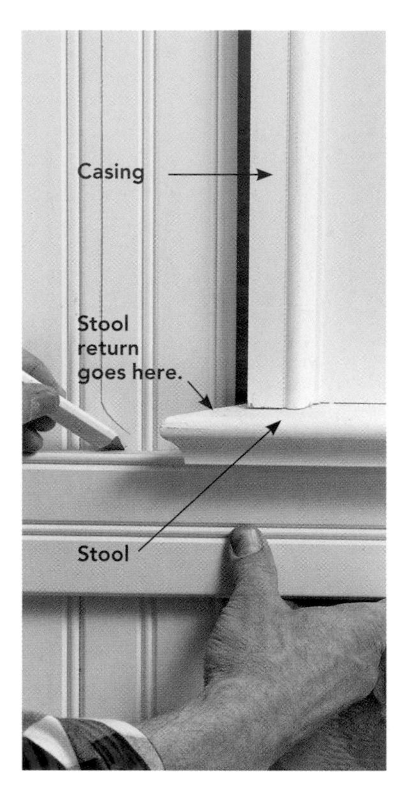

...THEN THE HORIZONTAL LINE. Holding the scribe guide against the bottom of the stool, mark the horizontal cut on the dry-fit board. The actual cut should be made about ⅛ in. below that line.

SLIP IN THE LAST BOARD. Install the scribed board against the window casing first, then slide the last full board into place from the top down.

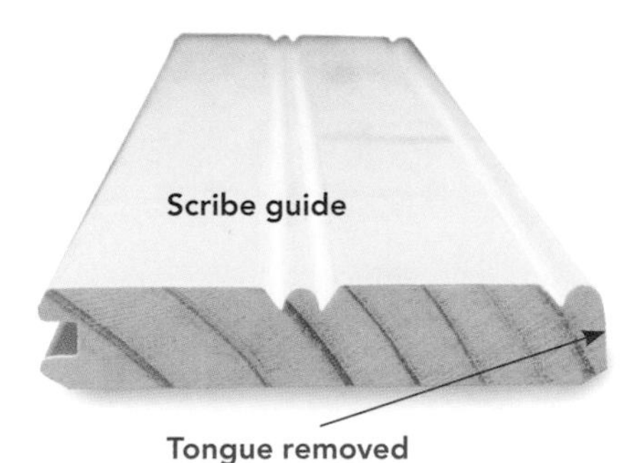

EXIT SIDE OF THE WINDOW

SCRIBE THE LINES. On the other side of the window, dry-fit an extra board under the window to space a full board properly for the scribe. Then use the scribe guide as before to mark the vertical and horizontal cutlines.

SCRIBED BOARD COMPLETES THE WRAP. After removing the extra dry-fit board, fasten the scribed board into position. Now you're ready to continue down the wall.

OTHER MATERIAL CHOICES FOR BEADBOARD

Over the years I have installed different types of beadboard. Some worked better in certain situations than others, and some I just should have avoided. A general rule of thumb is that the thinner the stock, the shallower the beaded profile. Shallower profiles fill with paint, and their shadowlines don't look nearly as nice as deeper beads milled in thicker stock.

MDF SHEETS Medium-density fiberboard (MDF) sheet-stock beadboard comes preprimed in 4-ft. widths. Take extra care to seal cut ends when installing it in moisture-prone areas such as bathrooms.

PLYWOOD SHEETS Usually ¼ in. thick, 4-ft.-wide plywood beadboard can be stained or painted. However, I've seen it split along a grain line and across bead lines occasionally, so it should be handled with care.

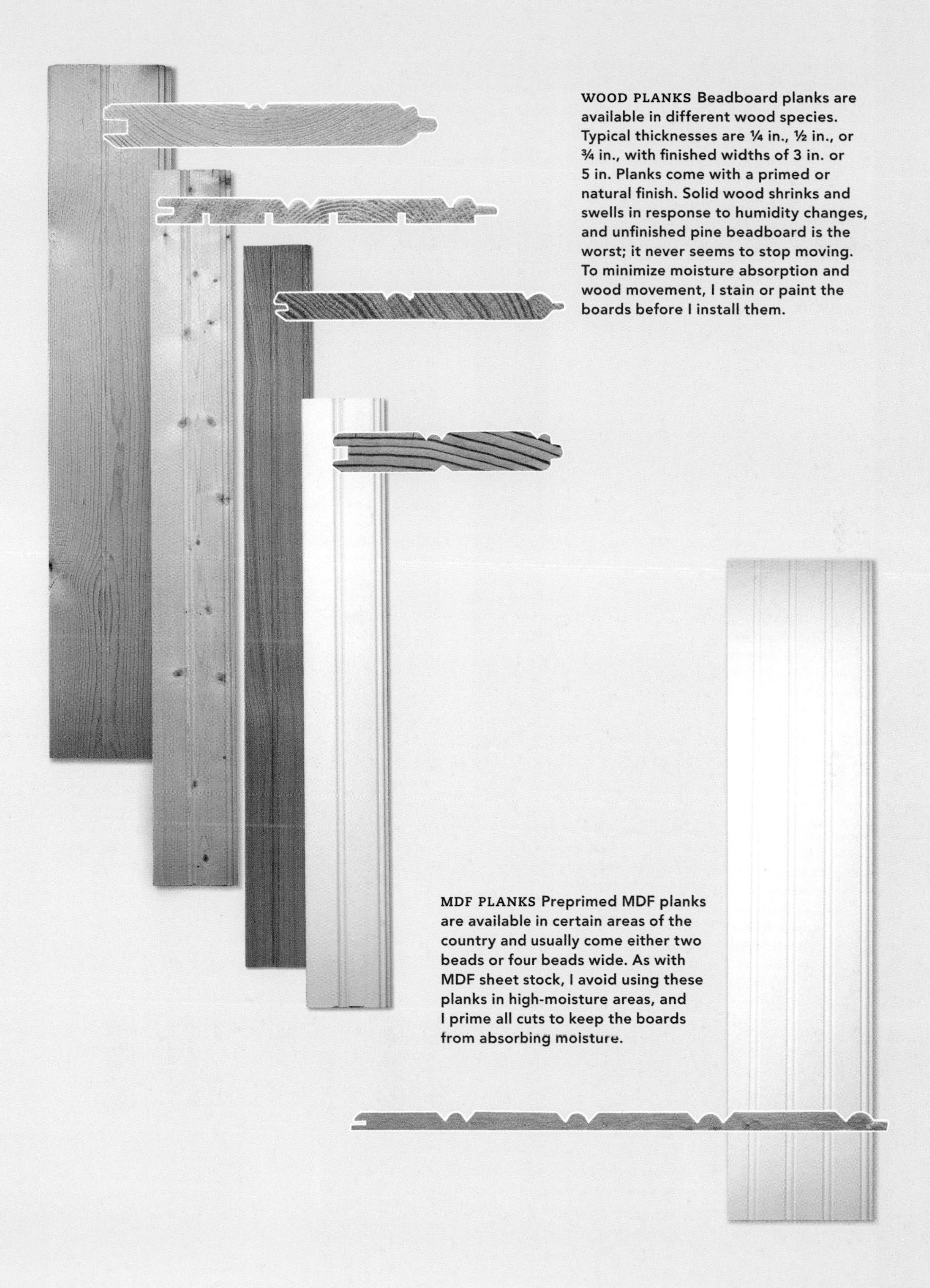

WOOD PLANKS Beadboard planks are available in different wood species. Typical thicknesses are ¼ in., ½ in., or ¾ in., with finished widths of 3 in. or 5 in. Planks come with a primed or natural finish. Solid wood shrinks and swells in response to humidity changes, and unfinished pine beadboard is the worst; it never seems to stop moving. To minimize moisture absorption and wood movement, I stain or paint the boards before I install them.

MDF PLANKS Preprimed MDF planks are available in certain areas of the country and usually come either two beads or four beads wide. As with MDF sheet stock, I avoid using these planks in high-moisture areas, and I prime all cuts to keep the boards from absorbing moisture.

Electrical boxes: Two ways to measure

The electrical code prohibits any combustible material, such as wooden beadboard, from being inside the area of an electrical box. When beadboard is installed after the electrical boxes, a box extension (see the top left photo) must be put in to avoid a code violation.

FOR ELECTRICAL BOXES THAT SHARE A SEAM BETWEEN BOARDS, simply dry-fit the first board, then mark and cut it accordingly. Once both sides are cut and checked for fit, fasten them to the wall.

FOR A BOX THAT FALLS WITHIN A SINGLE BOARD, mark the top and bottom on the plywood, then transfer the marks to the dry-fit board. Measure from the adjacent board for the vertical cuts.

Folded finish in a corner

I don't bother to fit the corner board precisely for the first wall because the edge is covered by the corner board from the other wall. To make sure that I don't end up with a narrow sliver in the corner, I take a quick measurement of the wall before installing any beadboard. If need be, I start with a half-width piece instead of a full board.

A SMART START. Narrow slivers of beadboard are difficult to install in a corner. To avoid them, a quick overall measurement can tell you whether to start at the casing with a full board or to rip the first board in half.

DRY-FIT FOR A PERFECT CORNER. After dry-fitting the last full-width board, push the scribe guide hard into the corner, and draw the cutline. Then cut the scribed board to become the final corner piece.

AN ACCORDION FIT. To complete the corner, pull the shared edge of the last full board and the scribed board out from the wall, seating their outside edges in place. Then spring the boards into place.

Wainscot for a Window

BY GARY STRIEGLER

Window-trim details can have a huge effect on the overall look of a room. With the right combination of materials and molding proportions, window-trim details transform a drab space into an elegant one. Getting those details wrong, however, really can disrupt the room's design.

Many of the houses I build have large windows with sills that are close to the floor. If I install a tall baseboard molding, I'm left with a strip of awkward-looking drywall beneath the window. To avoid that strip, I like to build a wainscot panel that extends from the windowsill to the baseboard molding. This detail grounds the window by connecting it to the baseboard and the floor. The window gains mass, and the little bump-out created by the wainscot has a big impact, breaking the wall plane with molding profiles that add visual interest.

MOLDING PROFILES DICTATE THE PROPORTIONS

This wainscot detail works well when these profile sizes are used on a window from 18 in. to 30 in. off the floor with 6-in.- to 7-in.-tall baseboard. Use casing widths ranging from 3 in. to 4½ in. Working outside these dimensions can make the field of the panel either too small or too large, which creates awkward proportions. Before you cut, take the time to plan the size of the wainscot panel carefully, and make a cutlist for the rails and stiles. Because the area behind the panel is going to be hidden, I lay out the dimensions right on the wall, sometimes going as far as drawing everything to scale if I'm concerned about how the overall proportions will look. This takes some time, but it lets me work out problems before any wood has been cut.

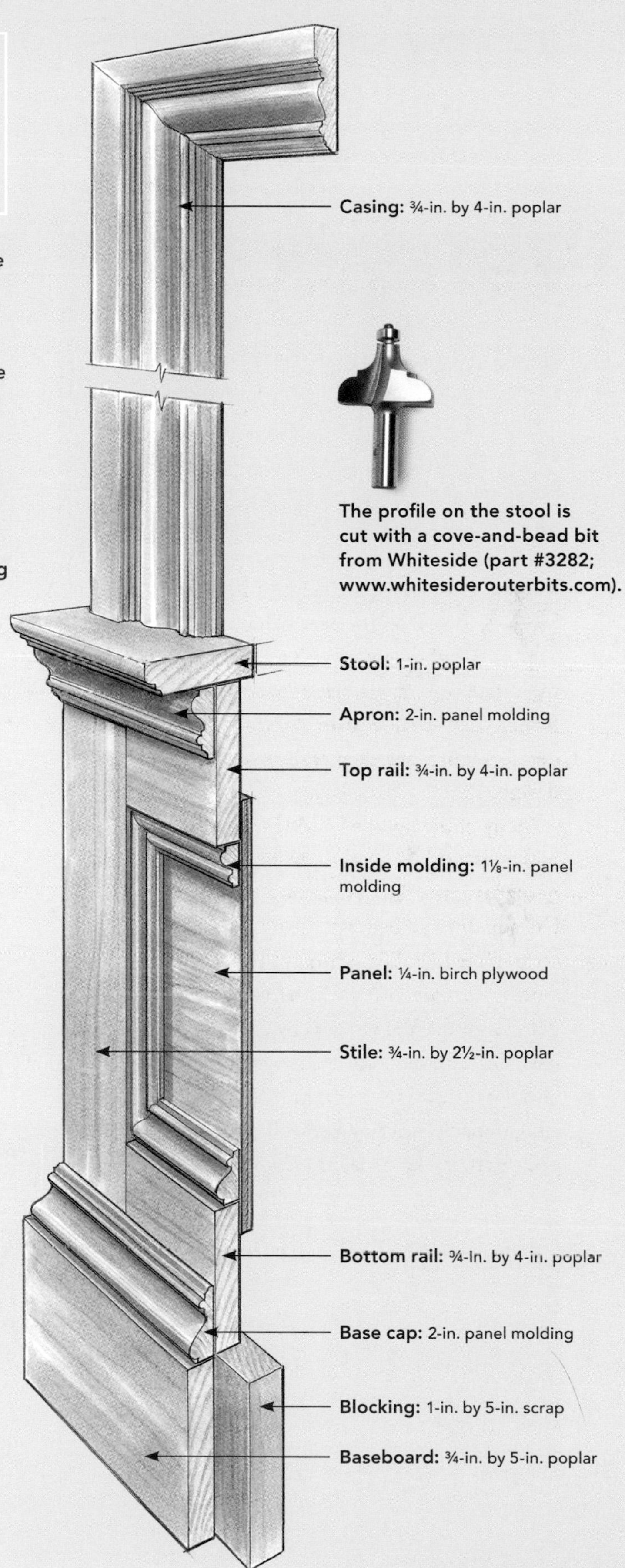

ASSEMBLE PIECES ON A WORKBENCH

With the rails and stiles cut, I assemble the frame with pocket-hole joinery, then add the plywood. During assembly, I make sure that the good side of the rails, stiles, and plywood will face outward when the panel is attached to the wall. I find it easier to build the frame and apply the stool, apron, and inside-panel moldings on a workbench before installing it.

1. JOIN THE FRAME WITH POCKET SCREWS. I bore two pocket holes at each stile-to-rail joint, spacing the holes about 4 in. apart along the top rail where the stool will be attached later. Before attaching the plywood back, I smooth the frame with an orbital sander.

2. ADD THE INSIDE MOLDING AFTER THE PLYWOOD IS ATTACHED. I glue and staple ¼-in. plywood directly to the back of the frame, keeping it clear of the pocket holes on the top rail. On the front, I install the short pieces of panel molding first, then nail the longer pieces to fit so that I can spring them in slightly for a tighter joint.

TIP: Back-bevel the miters with a block plane or a utility knife for a tighter joint.

3. **SCREW THE FRAME TO THE STOOL.** I mill the stool from 1-in. poplar with a cove-and-bead bit, then attach it to the frame with pocket-hole screws. I use a combination square to set the frame back ¼ in. from the stool's edge, the thickness of the plywood attached to the back of the frame. Kreg's® Right Angle Clamp has a foot on one end and a peg on the other to hold parts together until screws can be driven. (www.kregtool.com).

4. **ADD THE APRON NOW TO SAVE YOUR KNEES.** While the panel is upside down on the workbench, attach the apron molding and apron returns. Remember to run each return past the back of the frame ¼ in. so that it will meet the wall when it's installed.

Add the Final Moldings After the Panel is on the Wall

Once the frame is built and the moldings are applied, I attach the unit below the window, keeping the top of the stool ¼ in. to 3/16 in. beneath the top of the jamb to create a reveal. Blocking and baseboard moldings are next; then I install the casing around the window to complete the job.

5. **LEVEL AND CLAMP THE PANEL IN PLACE.** If the windows were installed properly, leveling the panel unit will keep the reveal between the stool and the jamb even. If the window isn't perfectly level, I hide the difference in the reveal before I nail the unit to the wall.

LOCATE BOXES WITH CENTERLINES. Plumb a line down the wall at the window's center and draw a line down the center of the back of the panel. Measure off these two lines and the top of the stool to pinpoint the cutout for an electrical box. An alternative approach is to have an electrician locate electrical boxes horizontally in the baseboard.

6. ADD BLOCKING AT STUD LOCATIONS TO SUPPORT THE BASEBOARD. Only a small part of the baseboard will overlap the bottom of the panel, so this backing provides nailing blocks for the baseboard.

7. NAIL THE BASE CAP TO THE PANEL AFTER THE BASEBOARD IS INSTALLED. I like to glue and nail the returns at the ends of the baseboard before attaching it to the wall. Once this is done and the base cap is on, I run a small cove strip to hide the ¼-in. gap left by the plywood panel. To eliminate this step, remove the drywall to allow the stiles to sit flush against the wall.

8. INSTALL THE HEAD CASING LAST. Nail the legs of the casing on first, keeping the same reveal between the casing and the jamb that was used between the stool and the jamb. Cut the head casing to fit, fine-tuning the angle as needed for a tight miter joint.

A Simple Approach to a Paneled Passageway

BY GARY STRIEGLER

Look in just about any old building, and you'll see that the paneled passageway has been around for a long time. This trimmed-out space has a dramatic impact on the overall look of a home's interior. Even more important, a passageway can add a sense of privacy to the room it serves. Adding this transition area can make a room feel more peaceful and sequestered. When I'm building a house, my clients often ask me to add paneled passageways to enhance privacy in certain rooms, especially master bedrooms and studies.

The good news is that you don't need to be a furniture maker to build paneled passageways, and they don't require expensive materials. I use simple pocket-hole joinery along with birch plywood, 1× poplar, and basic poplar moldings to build the panels. The best aspect of this approach is that everything is built off a prehung-door unit. With the door hinges already mortised, adding a paneled passageway to your next project requires only a little extra time and a few basic tools.

CREATE AN ELEGANT VESTIBULE by adding frame-and-panel jamb extensions.

IT ALL STARTS WITH A PREHUNG DOOR

THE FRAME-AND-PANEL assemblies that form the sides of the vestibule are actually extensions of the side jambs from a prehung door. After removing the door and the hinges, I rip the side jambs down to 2 in. wide. Then I set aside the side jambs and head jamb and use the door's dimensions to lay out the jamb extensions (see the details below). I build the side-jamb extensions, the top panel, and the filler panel, and then assemble them near the door opening.

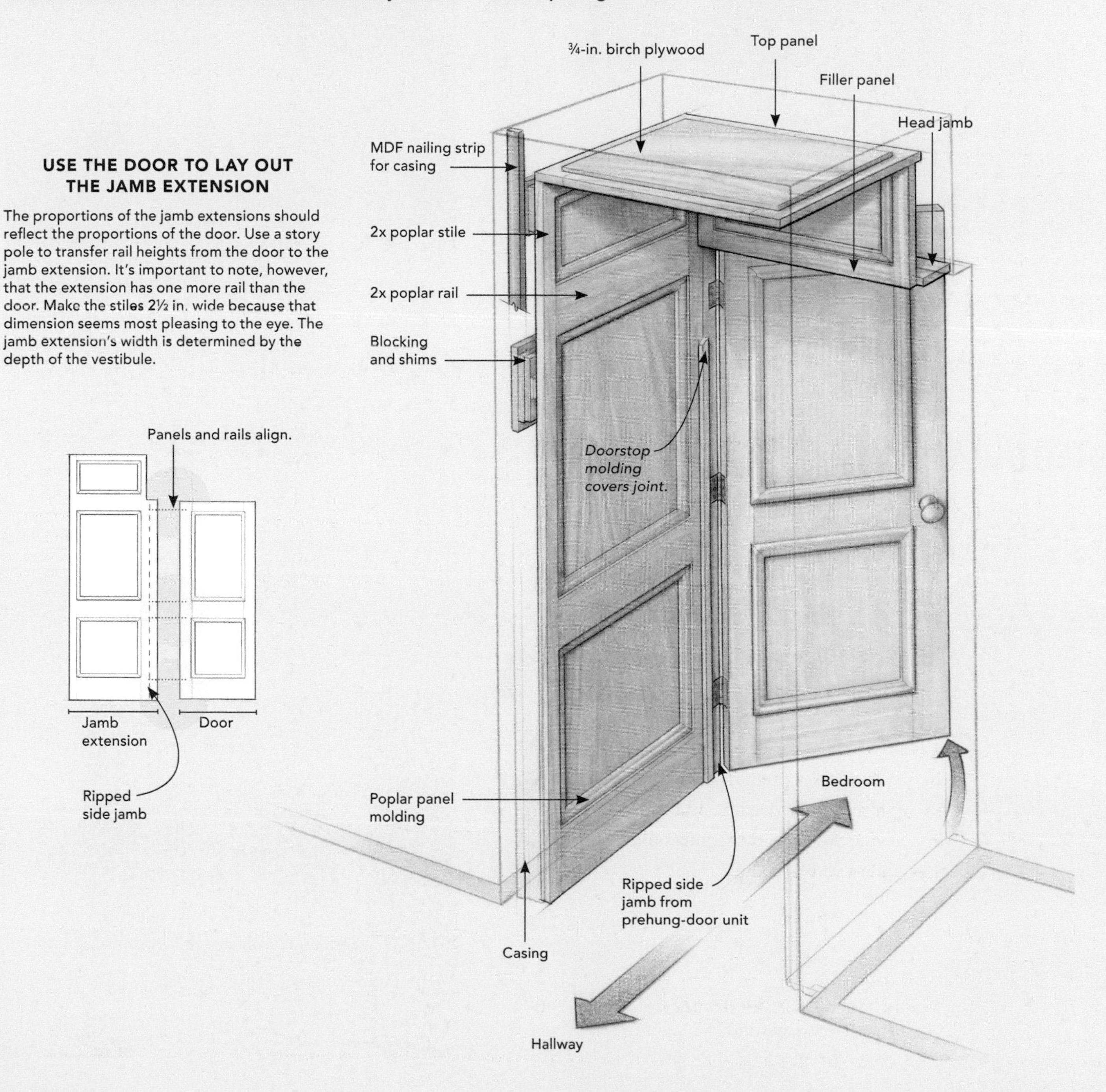

REMOVE THE DOORSTOP MOLDING, THEN RIP EACH SIDE JAMB. After disassembling the jamb, pry off the doorstop carefully so that it can be reinstalled later. Rip the side jamb down to 2 in. Also rip ¾ in. off the head jamb so that the filler panel will tuck in front of the head jamb when the unit is installed later.

A PASSAGEWAY CAN BENEFIT ANY ROOM

PANELED PASSAGEWAYS can do more than provide privacy. Adding a vestibule to a living-room entrance creates a niche for furniture [A]. Pushing a doorway into a room creates space for built-ins like desks and bookcases [B]. As shown in this article, vestibules also can incorporate existing partition walls from closets and bathrooms [C].

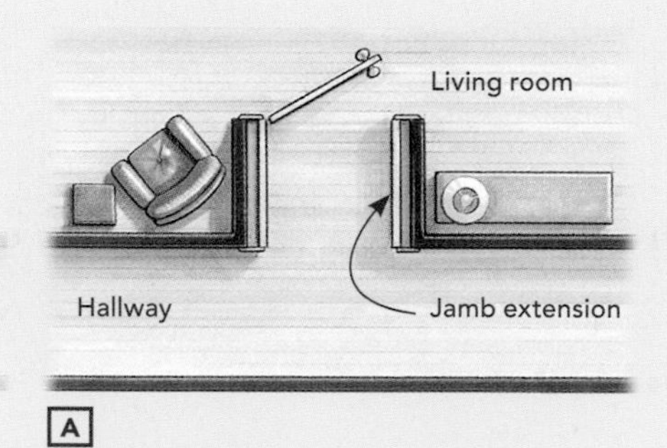

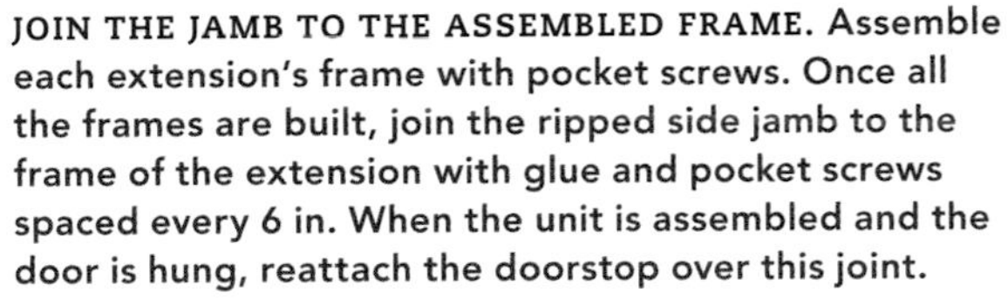

JOIN THE JAMB TO THE ASSEMBLED FRAME. Assemble each extension's frame with pocket screws. Once all the frames are built, join the ripped side jamb to the frame of the extension with glue and pocket screws spaced every 6 in. When the unit is assembled and the door is hung, reattach the doorstop over this joint.

ADD PLYWOOD AND PANEL MOLDING. After gluing and nailing ¾-in. birch plywood to the back of the panel frame, use a headless pinner to apply the poplar panel molding. All the vestibule panels are built the same way.

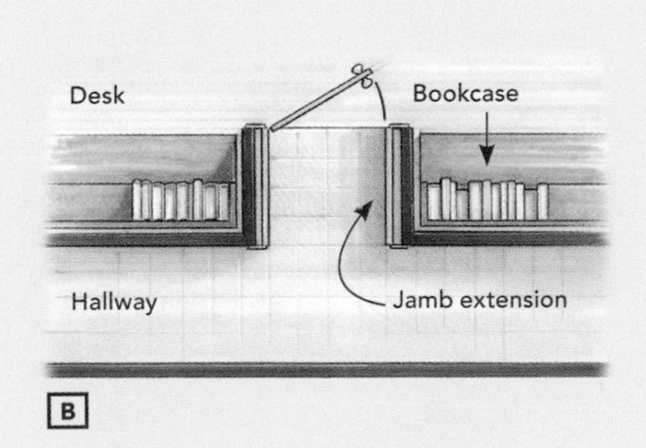

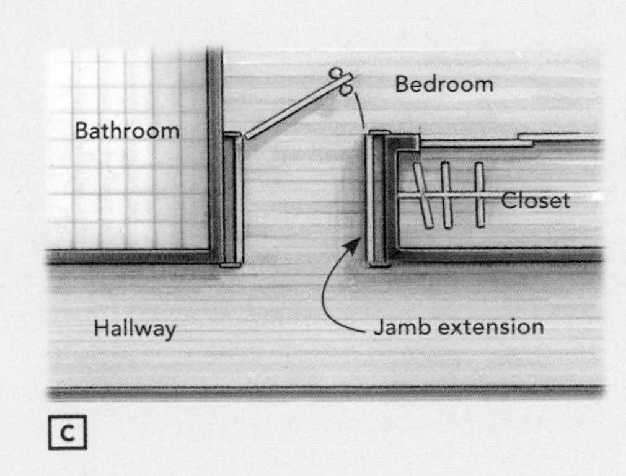

ASSEMBLE THE PIECES, AND HANG THE DOOR

The vestibule's side panels extend above the door's head jamb, which I install before tilting up the unit. Nailing up blocking ahead of time makes it possible to shim the panels plumb and level.

ATTACH THE TOP PANEL AND HEAD JAMB. Using the length of the original head jamb as a guide, screw the top panel and nail the original head jamb to the side panels. The author ripped ¾ in. off the original head jamb so that the filler panel will tuck in front of it once it's installed.

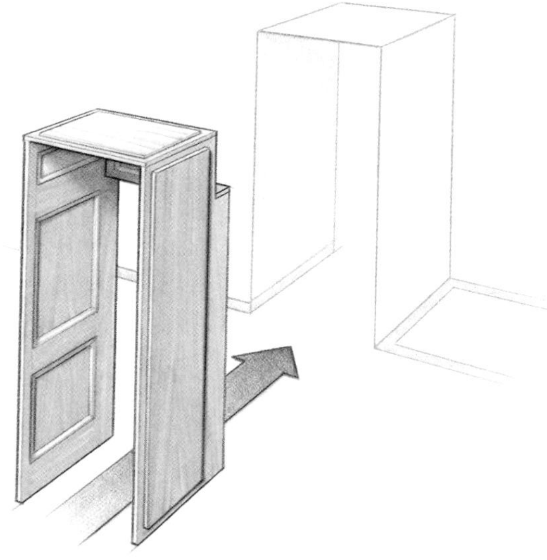

TILT AND SLIDE. After the original head jamb and top panel are attached to the side jambs, tip up the unit, center it in the rough opening, and slide it in place. An extra set of hands ensures that the unit won't rack as it's positioned.

SHIM THE STILES PLUMB, THEN ADD THE CASING

Once the door is hung, I finish hanging the rest of the unit. Although I could install the filler panel earlier, I choose to install it once the vestibule is nailed in place so that I can fine-tune the side jambs as needed. When everything is shimmed and nailed plumb, I reinstall the doorstop and nail up the casing.

SHIM THE JAMB PLUMB, THEN TACK IT IN PLACE. Plumb the hinge side of the unit first, then nail it in place with 2½-in. trim nails. Keep these nails in the area of the jamb that will be covered by the doorstop so that there are fewer holes to fill later.

INSTALL THE HINGES, THEN HANG THE DOOR. Once the hinge side of the jamb is nailed plumb, install the hinges. Whether hanging a hollow- or solid-core door, run a long wood screw through the middle hole of each hinge to catch the framing by at least 1 in. Then shim and nail the other side jamb in place, maintaining an even reveal around the door once it's closed.

SHIM AND NAIL THE OUTER STILES PLUMB. Before hanging the unit, the author installed blocking within shimming range (½ in. to 1 in.) to sit behind the extension's stiles. With the door jamb nailed off, insert and adjust shims to get the jamb extension plumb before nailing. A piece of medium-density fiberboard (MDF) provides backing for the casing.

SHIM THE STILES PLUMB, THEN ADD THE CASING (CONTINUED)

ATTACH THE FILLER PANEL TO THE HEAD JAMB. Nail the filler panel to the head jamb and to the studs, keeping the nails out of the panel's field. A small cove molding will cover the gap around three sides of the filler panel.

INSTALL THE DOORSTOP MOLDING AND CASING. With the door jamb and the door hung, reinstall the doorstop. The author likes to run the side pieces first and the header piece last. Finally, add the casing to each side of the opening.

CONTRIBUTOR LIST

8 Basic Rules to Master Trim Carpentry: Tucker Windover is a finish carpenter in Arlington, Mass.

Beautiful Trim from Plywood: Michael Standish is a carpenter living in West Roxbury, Mass.

Paint-Grade Interior Trim: Chris Ermides is an associate editor at *Fine Homebuilding*.

4 Router Tricks for Trim: Gary Striegler is a builder in Fayetteville, Ark.

Signature Trim Details: Charles Bickford is a senior editor at *Fine Homebuilding*.

Installing Baseboard: John Spier and his wife, Kerri, own Spier Construction, a custom-home building company on Block Island, R.I.

Baseboard Done Right: *Fine Homebuilding* contributing editor Gary M. Katz lives in Reseda, Calif. His website is www.garymkatz.com.

Crown-Molding Fundamentals: Clayton DeKorne is a carpenter and a writer in Burlington, VT. He is the author of *Trim Carpentry and Built-ins* (The Taunton Press, 2002), and the co-author of *For Pros By Pros: Finish Carpentry* (The Taunton Press, 2007).

Inside Crown Corners: Tucker Windover is a finish carpenter in Arlington, Mass.; Chris Whalen is a finish carpenter in Missoula, Mont.

Coping Moldings: Tom O'Brien is a carpenter in New Milford, Conn.

The Secret to Coping Crown Molding: Former millwright Bill Shaw is the inventor of the Copemaster, a production machine that he manufactures in Ridgefield, Conn.

Site-Made Moldings in a Pinch: Kit Camp (www.northparkwoodworks.com) is a carpenter and woodworker in San Diego, Calif.

Trimming Windows: Jim Blodgett owns A Small Woodworking Company in Roy, Wash.

The Only Way to Trim Exterior Windows: Mike Vacirca co-owns LastingNest Builders (www.lastingnest.biz), a building and remodeling company in Seattle, Wash.

Perfect Miter Joints Every Time: Jim Chestnut has more than 30 years experience as a carpenter. He lives near Bangor, ME.

Trim Windows with Built-Up Casings: *Fine Homebuilding* contributing editor Rick Arnold is a builder in North Kingstown, R.I.

Trimming a Basement Window: Chris Whalen is a partner in Black Mountain Company, a home-building, renovating, and woodworking firm in Missoula, Mont.

Craftsman-Style Casing: Tucker Windover is a finish carpenter in Arlington, Mass.

Replace an Old Entry Door: Emanuel Silva runs Silva Lightning Builders in North Andover, Mass.

Plumb Perfect Prehung Doors: Gary Striegler is a builder in Fayetteville, Ark.

Troubleshooting a Prehung Door Installation: Tucker Windover is a finish carpenter in Arlington, Mass.

No-Trim Door Jambs: Kevin Luddy owns Keltic Woodworking in Wellfleet, Mass.

Case a Door with Mitered Trim: Tom O'Brien is a carpenter in New Milford, Conn. Technical assistance by Tim Carney of Carney Home Enterprises in New Milford, Conn.

Turning Corners with Beaded Casing: Scott McBride is a contributing editor to *Fine Homebuilding* and the author of *Build Like a Pro: Windows and Doors* (The Taunton Press, 2002).

Add Elegance with Raised-Panel Wainscot: Gary Striegler is a builder in Fayetteville, Ark.

Paint-Grade Wainscot on a Curved Stairwell: Joe Milicia runs Hobart Builders in Fairfield, Conn.

Laying Out Wainscot Paneling: Architect Lynn Hopkins lives and practices in Lexington, Mass.

Designing with Wainscot: John S. Crowley, a former architecture professor, operates New England Classic® Inc. (www.newenglandclassic.com), an architectural-products manufacturer in Portland, ME.

Add Style with Beadboard Wainscot: *Fine Homebuilding* contributing editor Rick Arnold is a builder in North Kingstown, R.I.

Wainscot for a Window: Gary Striegler is a builder in Fayetteville, Ark.

A Simple Approach to a Paneled Passageway: Gary Striegler is a builder in Fayetteville, Ark.

CREDITS

All photos are courtesy of Fine Homebuilding magazine (FHB) © The Taunton Press, Inc., except as noted below:

Front cover: Photo by John Ross (FHB). Back cover: Photo by © James Kidd.

p. 2 Photo by Roe A. Osborn (FHB).

The articles in this book appeared in the following issues of Fine Homebuilding:

pp. 5–12: 8 Basic Rules to Master Trim Carpentry by Tucker Windover, issue 208. Photos by Charles Bickford (FHB), except photos on p. 5 and p. 7 (top right) by © Nat Rea.

pp. 13–17: 11 Beautiful Trim from Plywood by Michael Standish, issue 157. All photos by Chris Green (FHB). Drawings by Bob La Pointe (FHB).

pp. 18–27: Paint-Grade Interior Trim by Chris Ermides, issue 193. Photos by Krysta S. Doerfler (FHB) except photo on p. 22 by Charles Bickford (FHB); photos on p. 21 © courtesy Windsor Mills; photo on p. 24 (top left) © courtesy Pac Trim.

pp. 28–33: 4 Router Tricks for Trim by Gary Striegler, issue 198. All photos by Chris Ermides (FHB).

pp. 34–40: Signature Trim Details by Charles Bickford, issue 216. Photos by Charles Bickford (FHB) except photos on pp. 35 and 37 by © Brian Vanden Brink; photo on p. 40 (top left) by © Michael Mathers; photo on p. 40 (top right) by © Stephen Cridland. Drawings by Dan Thornton (FHB).

pp. 42–48: Installing Baseboard by John Spier, issue 146. All photos by Roe A. Osborn (FHB). Drawings by Dan Thornton (FHB).

pp. 49–58: Baseboard Done Right by Gary M. Katz, issue 174. All photos by Roe A. Osborn (FHB). Drawings by Martha Garstang Hill (FHB).

pp. 59–67: Crown-Molding Fundamentals by Clayton DeKorne, issue 152. Photos by © Andrew Kline except photos on p. 66 by Tom O'Brien (FHB). Drawings by Dan Thornton (FHB).

pp. 68–73: Inside Crown Corners by Tucker Windover, issue 209. All photos by Rob Yagid (FHB). Drawings by Dan Thornton (FHB).

pp. 74–76: Coping Moldings by Tom O'Brien, issue 164. All photos by Andy Engel (FHB).

pp. 77–79: The Secret to Coping Crown Molding by Tom O'Brien, issue 194. All photos by Charles Bickford (FHB).

pp. 80–83: Site-Made Moldings in a Pinch by Kit Camp, issue 206. Photos by Justin Fink (FHB) except photo on p. 81 (bottom) by Dan Thornton (FHB).

pp. 85–93: Trimming Windows by Jim Blodgett, issue 137. All photos by Andy Engel (FHB). Drawings by Dan Thornton (FHB).

pp. 94–99: The Only Way to Trim Exterior Windows by Mike Vacirca, issue 205. Photos by Chris Ermides (FHB) except photo on p. 96 by Rodney Diaz (FHB). Drawings by John Hartman (FHB).

pp. 100–107: Perfect Miter Joints Every Time by Jim Chestnut, issue 164. Photos by Randy O'Rourke except photo on p. 102 (bottom left) by Tom O'Brien (FHB) and p. 102 (right top and bottom) by Scott Phillips (FHB). Drawings by Dan Thornton (FHB).

pp. 108–115: Trim Windows with Buit-Up Casing by Rick Arnold, issue 172. All photos by Roe A. Osborn (FHB). Drawings by Chuck Lockhart (FHB).

pp. 116–125: Trimming a Basement Window by Chris Whalen, issue 189. All photos by Daniel S Morrison (FHB). Drawings by Bob La Pointe (FHB).

pp. 126–135: Craftsman-Style Casing Window by Tucker Windover, issue 196. All photos by John Ross (FHB). Drawings by Bob La Pointe (FHB).

pp. 136–141: Replace an Old Entry Door by Emanuel Silva, issue 218. Photos by Rob Yagid (FHB) except for photos on p. 139 (bottom right) by Dan Thornton (FHB). Drawings by John Hartman (FHB).

pp. 142–150: Plumb Perfect Prehung Doors by Gary Striegler, issue 202. All photos by Chris Ermides (FHB). Drawing by Dan Thornton (FHB).

pp. 151–157: Troubleshooting a Prehung Door Installation by Tucker Windover, issue 217. All photos by Charles Bickford (FHB).

pp. 158–163: No-Trim Door Jambs by Kevin Luddy, issue 153. All photos by Roe A. Osborn (FHB). Drawing by Dan Thornton (FHB).

pp. 164–167: Case a Door with Mitered Trim by Tom O'Brien, issue 169. All photos by Tom O'Brien (FHB).

pp. 168–175: Turning Corners with Beaded Casing by Scott McBride, issue 150. All photos by © Franklin Schmidt. Drawings by Vince Babak (FHB).

pp. 177–184: Add Elegance with Raised-Panel Wainscot by Gary Striegler, issue 165. Photos by © James Kidd (FHB) except photo on p. 184 (left) by Brian Pontolilo (FHB). Drawings by Bob La Pointe (FHB).

pp. 185–190: Paint-Grade Wainscot on a Curved Stairwell by Joe Milicia, issue 188. Photos by Charles Bickford (FHB) except photo on p. 187 (right) and p. 190 by Krysta S. Doerfler (FHB).

pp. 191–195: Laying Out Wainscot Paneling by Lynn Hopkins, issue 193. Drawings by Chuck Lockhart (FHB).

pp. 196–203: Designing with Wainscot by John S. Crowley, issue 152. Photo by Chales Bickford (FHB). Drawings by Mark Hannon (FHB).

pp. 204–213: Add Style with Beadboard Wainscot by Rick Arnold, issue 173. All photos by Roe A. Osborn (FHB). Drawings by Dan Thornton (FHB).

pp. 214–219: Wainscot for a Window by Gary Striegler, issue 180. All photos by Chris Ermides (FHB). Drawing by Dan Thornton (FHB).

pp. 220–226: A Simple Approach to a Paneled Passageway by Gary Striegler, issue 188. All photos by Chris Ermides (FHB). Drawings by Bob La Pointe (FHB).

INDEX